Governance Frameworks for Major Pu

Governance Frameworks for Major Public Projects

International Practices and Experiences

Edited by

Vedran Zerjav
Department of Industrial Economics and Technology Management
Norwegian University of Science and Technology
Trondheim, Norway

Morten Welde
Department of Civil and Environmental Engineering
Norwegian University of Science and Technology
Trondheim, Norway

Gro Holst Volden
Department of Civil and Environmental Engineering
Norwegian University of Science and Technology
Trondheim, Norway

WILEY

This edition first published 2026

Registered Offices
John Wiley & Sons, Inc., 111 River Street, Hoboken, NJ 07030, USA
John Wiley & Sons Ltd, New Era House, 8 Oldlands Way, Bognor Regis, West Sussex, PO22 9NQ, UK

For details of our global editorial offices, customer services, and more information about Wiley products visit us at www.wiley.com.

The manufacturer's authorized representative according to the EU General Product Safety Regulation is Wiley-VCH GmbH, Boschstr. 12, 69469 Weinheim, Germany, e-mail: Product_Safety@wiley.com.

Wiley also publishes its books in a variety of electronic formats and by print-on-demand. Some content that appears in standard print versions of this book may not be available in other formats.

Library of Congress Cataloging-in-Publication Data

Names: Zerjav, Vedran editor | Welde, Morten editor | Volden, Gro Holst, 1976- editor
Title: Governance frameworks for major public projects : international practices and experiences / Vedran Zerjav, Morten Welde, Gro Holst Volden.
Description: First edition. | Hoboken : Wiley, 2026. | Includes bibliographical references and index.
Identifiers: LCCN 2026004799 (print) | LCCN 2026004800 (ebook) | ISBN 9781394298525 cloth | ISBN 9781394298549 adobe pdf | ISBN 9781394298532 epub
Subjects: LCSH: Public investments – Management | Economic development projects – Finance | Project management | Public administration | LCGFT: Case studies
Classification: LCC HC79.P83 G68 2026 (print) | LCC HC79.P83 (ebook)
LC record available at https://lccn.loc.gov/2026004799
LC ebook record available at https://lccn.loc.gov/2026004800

Cover Design: Wiley
Cover Image: © JuliusGideon/stock.adobe.com

Set in 9.5/12.5pt STIXTwoText by Straive, Chennai, India

Contents

About the Editors

Vedran Zerjav is Professor of Project Management at NTNU with extensive international experience, having held full-time academic appointments in four European countries. He is passionate about working across disciplines and sectors, bridging academia, industry, and policy through active engagement in collaborative initiatives. His research broadly focuses on major projects, particularly in the domain of infrastructure, and he has served as Principal Investigator for projects funded by national research councils, professional bodies, and industry. Since 2020, he has served as Associate Editor for the *International Journal of Project Management*, the leading academic journal in the field.

Morten Welde is a senior researcher at the Concept Research Programme at the Norwegian University of Science and Technology (NTNU) in Trondheim, Norway. His research focuses on improving the performance of major public projects, with particular emphasis on both ex ante and ex post evaluations. He holds an MA in Transport Economics from the University of Leeds, UK, and a PhD from NTNU.

Gro Holst Volden is the Director of the Concept Research Programme at NTNU in Trondheim. Her main areas of expertise include project governance, public decision-making processes, and the evaluation of major public investments. Volden has extensive experience spanning academia, the consulting industry, and government administration in Norway. She holds a master's degree in economics and a PhD in project management.

Notes on Contributors

Dominic Ahiaga-Dagbui is Associate Head of School (Research) at Deakin University's School of Architecture and Built Environment. His research investigates governance, appraisal, and societal impacts of large-scale infrastructure projects, advancing cost-effective procurement, stakeholder engagement, and client competencies to mitigate cost overruns and optimize delivery of major infrastructure investments.

Javad Bakhshi serves as Subject Coordinator and Associate Lecturer at the University of Wollongong's School of Business. He holds a PhD in Project Management from the University of Sydney. His research interests focus on complexity, governance, digitalization, and project architecture. He also serves as an associate editor for the *Journal of Responsible Production and Consumption*.

Cristiana Belu Manescu is a Senior Economist at the European Commission, with experience in a broad range of areas including quality of public finances, macroeconomic forecasting, and sustainability. Cristiana holds a PhD from the University of Gothenburg, Sweden, and has published work on public investment management, fiscal rules, oil markets, and corporate social responsibility.

Maude Brunet is Associate Professor in the Department of Management at HEC Montréal. Her research interests include project organizing, governance and innovation of megaprojects, public infrastructure projects, and public–private partnerships. She has published in several project management and administrative science journals. She is Associate Editor for the *European Management Review* (area of Project Management Studies).

Olivier Choinière, PhD, is Associate Professor of Project Management at the Université du Québec à Rimouski (UQAR) and affiliated researcher with the Centre on Governance at the University of Ottawa and the Conference of Defence Associations Institute. His research focuses on new approaches to project management, governance of military and public infrastructure projects, and the future of work in the public sector.

Ana Maria Dobre is Acting Head of Unit at the European Commission, specializing in employment, social affairs, and inclusion. With a PhD in Social Sciences from Leuven University and a Master's from the College of Europe, she has extensive experience in policy development, economic transitions, and regional development. Ana is known for her expertise in cohesion policy and her leadership roles in various European initiatives.

Garret Doocey is an Assistant Secretary General in the Irish Ministry of Transport with responsibility for Land Transport Investment. He has previously held roles in various different areas of the Ministry and began his Civil Service career in the Courts Service. Prior to joining the Civil Service, he worked in the private sector and holds an MSc in Major Programme Management, an MSc in Economic Policy Studies, and an MA in Public Management.

Nathalie Drouin, full professor in Project Management, Department of Management, ESG UQAM, and Adjunct Professor at University of Technology Sydney (UTS), is the Chairholder of the Research Chair INFRA-S on Social Value of Infrastructures and Editor-in-Chief of *IJMPB*. Her research looks at governance and social value of infrastructure projects, organizational project management, and megaprojects. She serves on several boards and has received many prestigious research awards.

George Edward Gibson is currently President and Chief Executive Officer of the National Academy of Construction (NAC). He is an expert in front-end planning, risk management, and leadership, having held a variety of successful academic and industry positions in his career, while consulting with over 80 organizations globally.

Jon Frazer has over 15 years of experience in public infrastructure policy and governance, having worked in senior roles across the Australian and New South Wales Governments, at Australia's leading infrastructure think-tank, and advised clients across the public and private sectors.

Andreas Claus Hansen has over 15 years of experience with the management of public construction projects. He has held leadership roles in project management, project cost teams, and consulting, including positions as Professional Lead, Head of Department and Head of Risk Management, demonstrating expertise in engineering project and cost management for large public construction client organizations.

Ed Hearne is a Senior Economist in the Fiscal Affairs Department of the IMF, providing advice on issues of public financial management and infrastructure governance to countries in Africa, Europe, the Middle East, and Central Asia. Prior to joining the IMF, Ed held senior roles in the civil service of Ireland. He is a graduate of University College Dublin and the University of Oxford.

Richard Kirkham is a Reader in Civil Engineering at the University of Manchester where he leads the SALIENT Research and Coordination Hub, a five-year government-funded research programme, which aims to address national security and resilience challenges through an inter-disciplinary approach. Prior to SALIENT, Richard was Deputy Director of the Thomas Ashton Institute for Risk and Regulatory Research, where he worked on the COVID-19 core national studies programme. He was seconded into the Cabinet Office in 2016, where he helped establish 'Project X,' and academic-civil service collaboration into major project delivery in government.

Niek Mouter is an Associate Professor of Government Project Appraisal who works at Delft University of Technology. He is an expert in Cost–Benefit Analysis, and he developed the Participatory Value Evaluation, which is used more than 100 times in real-life decision-making processes in the Netherlands.

Roger Pyddoke is a senior researcher at the Swedish National Road and Transport Research Institute VTI. By education, he is an economist from Stockholm University. He has worked with the Swedish Agency for Public Management (Statskontoret) and Swedish Institute for Communications Analysis (SIKA) before joining VTI. He has researched many aspects of transport policy.

Per Svejvig received the PhD degree from Aarhus University, Aarhus, Denmark, in 2010. He is currently an Associate Professor in Project Management with the Department of Management, Aarhus University. He has authored or co-authored articles in various journals, including the *IEEE Transactions on Engineering Management*, *Journal of Information Technology*, and *International Journal of Project Management*.

Terry Williams was in the defence industry for nine years. He has worked in three business schools: Strathclyde, Director at Southampton, and Dean at Hull. He has held various consultancy contracts, including supporting multiple $billion post-project arbitrations. He has held multiple research contracts, including on the UK 'Project X', and multiple PMI-funded projects. He sits on the UK's REF panels and PMI's Academic-Insight-Team. He has authored >100 journal articles (h-index 58) and various books.

Nicola Young is a UK Government civil servant and, for nearly 30 years, has worked across a range of project specialisms. She is a Chartered Project Professional and an honorary visiting researcher for the UoM Thomas Ashton Institute, providing both practitioner and academic insight. Her research interests include project governance, assurance, and delivery methodologies.

Foreword by

Terri Harrington is a semi-retired project professional who, until March 2023, was a Sponsorship Director in National Highways with accountability for >£20 billion of major road projects. In this role, she focussed on governance, relationships with central government, securing investment, and assurance activities. Previously, Terri was the Head of Portfolio Insight at the IPA (now NISTA) and in this role collected, collated, analysed, and reported on data from projects on the Governments Major Project Portfolio. Initially using GMPP data, she also set up and led Project X, a unique collaboration between academics, government, and professional bodies, which sponsored practical-based research to improve project delivery outcomes.

Introductory Note from the Editors

This book started with a simple idea back in late 2023: update a study from 2016 – *Concept Research Report 46–47* – that looked at how different countries approach project governance. The plan was to write a technical report, publish it on our website, and move on.

But as we got into it, the project took on a life of its own. With support and enthusiasm from the broader Concept Research group, we decided to do things a bit differently. We reached out through our networks and brought together a group of international experts who live and breathe project governance. The goal was not just to collect insights, but it was also to learn from each other and build something together.

Over the next year and a half, we met regularly – not just to coordinate but also to have real, open conversations. We shared experiences, explored challenges, and compared how governance works across different countries and sectors. We wanted this to be a team effort, and it really was.

We also held two workshops at our NTNU Brussels office. Those two-day sessions were some of the highlights of the whole process. They gave us space to dig deeper, challenge each other's thinking, and build a shared understanding. What came out of those meetings was not just content for a book, but it was also a sense of community.

And speaking of community – we are incredibly grateful to the fantastic team of contributors who made this journey not only insightful but also genuinely enjoyable. It was not just about the delicious baguettes and dinners in Brussels restaurants (though those were great), but it was also about the deep, inspiring conversations that happened around those tables. Together, we created a network that brought together people from sectors and countries that barely ever get to talk. Thanks to you, we had conversations that made us richer as experts, professionals, and as people.

So while this book is the visible result of our work, it is only part of the story. Behind it is a process of learning, trust-building, and collaboration across sectors and borders. We hope the book reflects that spirit – not just in what it says but also in how it came to be.

There is still more to explore – more countries, more types of projects – but we see this as a strong starting point. Whether you are working in policy, managing projects, or just curious about how governance works in different contexts, we hope you find something useful here.

Thanks for picking up the book. We hope it sparks ideas, conversations, and maybe even new collaborations.

Trondheim *The Editorial Team*

Foreword

I was delighted to be asked to review this book 'Governance Frameworks for Major Public Projects: International Practices and Experiences' and provide a Foreword to this book.

The book was co-authored by an international group of academics and project practitioners. In my experience of leading Project X[1] for many years in the United Kingdom, project practitioners and academics working together blend different skills, knowledge, and experience and provide unique insights, which can help improve investment and delivery outcomes.

It successfully achieves this by providing evidence-based analysis of the effectiveness of a range of governance frameworks and practices, and therefore, the book is of considerable value to academics, policy makers, and project practitioners wherever they operate.

In Parts I and II, the book compares governance frameworks and practice in nine countries (plus the World Bank and European Commission) and identifies a range of opportunities and pitfalls, which could impact on the success of project delivery anywhere.

In addition to developing an understanding of the operation of governance in multiple contexts, by reading this book, academics will learn more about the effectiveness of cross-border and cross-disciplinary research collaboration in bringing together a comprehensive and useful body of knowledge.

The comparison of how different types of projects are governed in a range of contexts will support policy makers in reaching evidence-based decisions on how to establish the governance of new change initiatives.

Part III of the book identifies the common factors influencing the impact of governance frameworks in different contexts and the potential pitfalls to avoid. Project practitioners, from either the public or private sector, can apply this learning to their projects to minimize the risks identified and optimize the probability of successful delivery.

The book stands alone but also provides a future foundation that could be developed further into a summary of best-practice recommendations arising from the research. If presented in such a way as to be widely accessible to project practitioners and others, this could have a further significant and positive impact on future project delivery outcomes globally.

1 Project X was a collaboration between the Infrastructure and Projects Authority (part of UK Cabinet Office and HM Treasury) (now NISTA), academics, and professional bodies developing evidence-based research aimed at improving project delivery outcomes in the United Kingdom.

I would like to thank the editors and authors of the book for collating a comprehensive and extremely interesting set of research findings. This was a significant undertaking, and bringing together the outcomes of the analysis of the governance frameworks, practices, successes, and pitfalls in so many different contexts is unique. It also provides an excellent starting point for further developing the application of the findings for policy makers and practitioners and identifies potential areas for future research for academics.

As a long-standing project practitioner in the United Kingdom, I found the book fascinating and believe that the insights it provides shine a light on how to improve the effectiveness of governance of major projects globally.

Terri Harrington
London

Generative AI Disclosure

In writing the book, we utilized the support of generative AI (Copilot and ChatGPT-5) to develop early outlines and draft minor parts of the introduction and conclusion chapters. This was based on summaries of our original meeting notes, which were then revised manually based on our subject matter expertise. Similarly, generative AI was used in minor editorial interventions in the individual country chapters, for drafting the chapter preface paragraphs for several of the country cases. We used Grammarly to improve the quality of language and to streamline the argument for flow and coherence. All AI-generated content was comprehensively reviewed and revised, and we take full editorial responsibility for the content in this book. Authors have made separate disclosures about the use of generative AI in individual chapters.

Acknowledgements

We acknowledge the Norwegian Ministry of Finance for its continuous support of the Concept Research Programme and for providing funding for the project leading to this book. We gratefully acknowledge Terri Harrington for her enthusiasm to engage with the draft and write the Foreword at the 11th hour. Many thanks to our old friend and collaborator, Terry Williams, for serving as the 'shadow senior editor-in-chief', being a continuous voice of support and a friendly critic, and for ensuring we remain focused and on track to achieve our goals while avoiding unnecessary distractions.

Part I

Introduction to Governance of Public Investment Projects

1

Introduction – Governance of Public Investment Projects

Vedran Zerjav, Gro Holst Volden and Morten Welde

Concept Research Programme, Norwegian University of Science and Technology (NTNU), Trondheim, Norway

1.1 Governance Arrangements for Project Development and Implementation

Major public investment projects (sometimes referred to as major government projects) are one of the key mechanisms for the implementation of government policies. Such projects can be found in various sectors, including infrastructure (transport, information, and communication technology [ICT], water, energy), defence, systems transformation, or organizational change, and they are often not only significant in terms of scale but also complex and politically significant. Major government projects must deliver on policy priorities and societal needs, but they have a poor track record of delivery. Cost overruns, delays, and failure to meet intended objectives are common, and this book aims to address some of these challenges from the perspective of project governance.

Governance, especially at the front-end of projects, has long been recognized as a key determinant of successful public investment projects. Project governance involves establishing and implementing arrangements, structures, and processes to facilitate decision-making, establish accountability, and ultimately align project delivery with the policy goals they are intended to deliver on (Samset and Volden 2016). Recently, a growing body of literature on project governance has emerged, accompanied by an increasing recognition of the role of governance in shaping long-term project outcomes. Nonetheless, much of this literature remains primarily directed at academic audiences.

Over the past two decades, several countries have also introduced practical project governance arrangements aimed at ensuring that large public investment projects are justified, well planned, and efficiently delivered. Frameworks from six countries were presented in a study by the Concept research programme 10 years ago (Samset et al. 2016, later published as an academic paper, Volden and Samset 2017). Many more schemes exist. They share similarities and differences, and together, they represent a wealth of experience that should be shared across countries.

This book aims to bridge the gap between policy intentions and project delivery. It builds on Samset et al. (2016) but expands the scope by including more countries

Governance Frameworks for Major Public Projects: International Practices and Experiences, First Edition.
Edited by Vedran Zerjav, Morten Welde and Gro Holst Volden.

and their respective schemes, as well as more recent experiences. In contrast to the 2016 study – which was based on a desk review conducted by Norwegian researchers – this book is the result of a collaborative project involving experts from all the included countries. These contributors bring valuable, close-up knowledge of the schemes, enriching the analysis with first-hand insights. The book aims to bring recent research on project governance into the realm of policy and industry practitioners – those responsible for planning, sponsoring, and delivering major public investments. These practitioners often operate under time constraints and require insights that are both theoretically grounded and practically applicable. In this book, we aim to provide such applicable insights, informed by empirical research and experience across different countries.

The editorial team's involvement in initiatives such as the Concept Research Programme in Norway and Project X in the United Kingdom has highlighted the importance of governance in improving project outcomes. These research initiatives have contributed to the ongoing development of governance models in dialogue with public sector project delivery practitioners. Drawing on this experience, the book offers a comparative perspective on governance arrangements across different country contexts to inform policy practice.

1.2 Governance Theory – Conceptual Foundations

This book focuses on the formalized governance frameworks introduced by the topmost level of government, that is the institutional frameworks and decision-making processes that guide project development and implementation. It involves the allocation of roles and responsibilities, the establishment of procedures for quality assurance (QA), and the creation of mechanisms for oversight and accountability.

Recent academic work has emphasized the importance of governance, particularly at the front-end of projects. This fuzzy phase begins with a vague idea about the problem to be solved and its solution (presumably in the form of a temporary project), and it ends with the decision to sanction the project for execution. Research suggests that this front-end phase of projects is essential to establishing the direction and long-term outcomes for the project. Similarly, strong governance at the front end can mitigate issues such as selecting the wrong projects for execution, biased cost estimates, and inadequate stakeholder engagement.

The literature on project governance is burgeoning, with more recent invaluable contributions from project management scholarship such as Müller et al. (2023), Williams et al. (2022), and Volden and Samset (2017). Such contributions (as elaborated upon further in Chapter 2) provide a theoretical and conceptual foundation for empirical analyses that highlight the role of governance, both at the front end of projects and during execution and handover. Governance also contributes to a more structured decision-making process, the delineation of organizational and institutional roles, all leading to stronger QA mechanisms.

Despite these developments, a significant gap remains in the literature that translates governance theory into practical guidance for policy and industry practitioners. With this book, we aim to address this gap by discussing and elaborating on formal project governance schemes in a format that is more accessible to project professionals, regardless of whether they are government officials or industry practitioners.

1.3 Rationale – Selection of Case Settings

The rationale for this book is based on the need to provide a comprehensive and comparative analysis of governance arrangements for major public investment projects. Our international experience has convinced us of the value of institutionalized governance models for delivering projects and improving their outcomes. These governance models often include stage-gates as a structure for decision-making to establish accountability, efficiency, and legitimacy.

In Norway, the Concept Research Programme focuses on the front-end governance of major public investment projects under the Norwegian State Project Model, also known as the QA scheme. This model, which was introduced by the Ministry of Finance in 2000 and applies to investments in all sectors, involves a stage-gate process with defined decision points, institutional roles, and QA requirements. The research programme has generated extensive empirical data and theoretical insights into how sound decisions can be made and projects effectively governed. The long-term ambition of the collaborative programme is to contribute to research, capacity building, and institutional reform, enabling an ongoing dialogue with the public sector project delivery bodies.

Some other countries have introduced similar research initiatives related to public project governance. In the United Kingdom, Project X – Improving Project Delivery – research initiative, which ran between 2018 and 2025, in collaboration with the Infrastructure and Projects Authority (IPA) – now renamed as the National Infrastructure and Service Transformation Authority (NISTA) – aimed to improve the governance of planning and delivery of the Government Major Project Portfolio (GMPP). In Denmark, a research network called Program Denmark was established in 2023 to improve the governance, planning, and execution of major public projects.

Building on experiences from these and other research activities in the area of project governance, the book includes country cases from Norway, Sweden, Denmark, the United Kingdom, Ireland, the Netherlands, Canada (both the Province of Quebec and the federal level), the United States, and Australia. While many other countries could have been taken to study project governance and its implementation, these countries were selected based on their relatively coherent use of formalized stage-gate models, institutional diversity, policy relevance, and the availability of experiences with their implementation. Our approach was to engage with a selected group of international experts from academia, civil service, or multilateral bodies; to achieve deeper engagement with the practices; and to strengthen the credibility of the analysis and discussion.

The country cases follow a consistent descriptive structure, enabling comparative analysis across key dimensions:

- Country context and background for introducing public project governance arrangements

- Presentation of the governance frameworks and stage-gate models, including institutional roles and responsibilities, QA mechanisms, etc.
- Experiences with the frameworks, including implications for project selection, delivery, and accountability

In addition to national-level examples, the book also briefly touches on critical perspectives from supranational and multilateral bodies, including the European Commission and the World Bank. They provide valuable international perspectives on governance practices and policy implementation, including the focus on the delegation of roles within the European Union (EU) budget implementation system and the separation of roles embedded in the no-objection principle for the approval of World Bank projects.

The comparative analysis intends to identify commonalities and differences in governance across countries and settings and to generate practical insights and recommendations for governments and other bodies working with project governance schemes or planning to implement them. Using a range of examples and experiences, we intended to develop a valuable resource for practitioners engaged in the planning and delivery of major public investment projects.

1.4 How to Use This Book

This book is structured to serve as both a reference work and a practical guide for government officials and industry practitioners tasked with implementing policies through projects. It comprises two main parts. After the introduction and a summary of key theoretical concepts in Part I of the book, we present detailed case studies of governance models in the selected countries and supranational/multinational organizations in Part II of the book (Chapters 3–14). Each chapter follows a structure that facilitates comparative analysis and discussion. The country cases provide insights into the design and operation of governance mechanisms, institutional roles, QA procedures, and the historical evolution of governance schemes.

The final section and Part III of the book (Chapters 15 and 16) develop a comparative analysis of the governance arrangements presented in the country cases. The analysis explores the relationship between project delivery, the project front-end, and policy intentions. The final segment also provides practical advice for institutional design and reform, addressing issues such as QA, culture and governmentality, and the roles of planners, project managers, sponsors, and owners. We conclude with a brief discussion of the contributions to knowledge on project governance and relate key findings with the key governance theoretical frameworks: Transaction Cost Economics (TCE), Agency Theory (AT), and Stewardship Theory (ST).

1.5 Intended Audience and Use

The book is written for government officials and industry practitioners who are responsible for the planning and implementation of major government investment projects.

These include civil servants, project sponsors, planners, and delivery managers. The book is also relevant for post-experience degree-level programmes (e.g. MSc, MBA, MPM, and MPA) and professional development courses focused on major projects and programmes, infrastructure, and policy implementation.

While the book can be read sequentially, it is designed for flexible use, allowing readers to focus on sections most relevant to their professional needs.

References

Müller, R., Sankaran, S., and Drouin, N. (2023). *Research Handbook on the Governance of Projects*. Cheltenham: Edward Elgar.

Samset, K. and Volden, G.H. (2016). Front-end definition of projects: ten paradoxes and some reflections regarding project management and project governance. *International Journal of Project Management* 34 (2): 297–313.

Samset, K., Volden, G.H., Olsson, N., and Kvalheim, E.V. (2016). Governance Schemes for Major Public Investment Projects. A comparative study of principles and practices in six countries, Concept report no 47, Ex ante Academic Publisher.

Volden, G.H. and Samset, K. (2017). Governance of major public investment projects: principles and practices in six countries. *Project Management Journal* 48 (3): 90–108.

Williams, T.M., Samset, K., and Volden, G.H. (ed.) (2022). *The front-end of large public projects: paradoxes and ways ahead*. Abingdon: Routledge/Taylor & Francis.

2

Introductory Concepts

Terry Williams

Hull University Business School, University of Hull, Hull, UK

2.1 Introduction

The aim of this chapter is to introduce some of the concepts that underlie the challenges described in this book: some of the key theoretical ideas and the challenges of implementation. This chapter looks at what governance means in the public domain, how a government governs, and how it ensures that public policies are designed and implemented in such a way as to satisfy what it intends those policies to achieve. Since much of government business is implemented through projects (and this is the subject of this book), the chapter looks at the nature of public projects. After reviewing the life cycle of projects, the chapter looks at governance of individual projects, and finally, the target of this book: governance frameworks for projects within the public sector, including a mention of how practical politics can interfere or interact with these frameworks. Some discussion is given of the object of these frameworks and their assessment, as well as some theoretical concepts to help understand them, in preparation for the following chapters of this book.

2.2 Public Governance

The word 'governance' refers to the processes by which any organization, such as a government or a corporation, governs those within its remit. This book is concerned with public governance, or governance exercised by a national government (although we also refer to a few supra-national organizations in this book); public governance, of course, has its counterpart in the private sector, namely corporate governance (see Tricker 2015), whence much of the relevant literature on project governance comes.

However, there are some distinct differences in public governance, clearest amongst them being that the relevant organizations are publicly owned; they are carrying out activity (e.g. providing public services) that is generally not in a free market; their incentives are not generally profit-maximization but might be multi-faceted, but also outside

Governance Frameworks for Major Public Projects: International Practices and Experiences, First Edition.
Edited by Vedran Zerjav, Morten Welde and Gro Holst Volden.

stakeholders might have wide and heterogeneous motivations, possibly inimical to the public body; furthermore, they might be under different legislation. These differences are explored more widely in Müller et al. (2023).

The growth of the term 'governance' reflects the changing nature of our public domain in recent decades, particularly with the growth of New Public Management (NPM) and its more recent post-NPM relatives such as New Public Governance (Osborne 2010). The main development was in the direction from an emphasis on competitiveness, efficiency, and value-for-money in NPM to the need for a broader consideration of public value for different stakeholders and the broader society (Osborne et al. 2021). In this way, 'governance can refer abstractly to all processes of government. It supplements a focus on the formal institutions of government with recognition of more diverse activities that blur the boundary of state and society. It draws attention to the complex processes and interactions involved in governing. Governance can also refer, more concretely, to the rise of new processes of governing that are hybrid and multi-jurisdictional with plural stakeholders working together in networks' (Bevir 2012, p. 5).

The discussion in this book will thus cover a wide range of frameworks, processes, and structures; these will be underpinned by policy instruments, be they regulation (affirmative or negative), economic means (affirmative or negative) or information (affirmative or negative), as described in Bemelmans-Videc et al. (1998), applied to public sector projects in Samset et al. (2016). For the purposes of this chapter, we shall refer to a framework for governance, or a regime, or a scheme; or a model; we will not specify a particular term.

For any sort of corporate governance, there are fundamental principles of what 'good' governance looks like, as espoused in a 1998 OECD report and discussed in Cadbury (2000): the key principles being

- **Fairness:** The fair and equal treatment of the wide variety of stakeholders
- **Transparency:** Disclosure of relevant information
- **Accountability:** Clear roles and responsibilities for the various parties involved
- **Responsibility:** Socially responsible behaviour.

These are particularly applicable to public governance and will be referred to in this book.

There is another word that will be necessary in this book: governmentality (originally expressed by Foucault 2007), 'a fusion of government and mentality', which means 'actively governing through mentalities' (Clegg 2019). As a soft factor, governmentality provides an integrative mechanism to observe and make sense of how governance, a hard factor, takes place in projects (Clegg et al. 2002). To reduce ambiguity, we will follow the definitions by Müller et al. (2023): governmentality is 'the study of the complex relationship between the subject and power along with their political rationalities, motivations and technologies through which governance occurs' (p. 79). They distinguish between four concepts on the steering and executing levels of government dealing with structure and with human agency: at the steering level, governance gives the structure and governmentality gives the 'governors' particular way of interacting with those they govern'; at the executing level, management gives the structure and goal-oriented activity, and leadership the people-oriented activity (Müller et al. 2023, p. 4).

2.3 Public Projects

The topic of this book covers major public projects. Why are these carried out? Governments make policy, but this policy is generally actually delivered through projects and programmes. The successful delivery of these projects is essential to achieve the various strategic and tactical objectives of the government, not only clear physical assets such as new infrastructure or military capability but also the more strategic but less obviously tangible objectives such as improving the efficiency or transparency of public services and sustainable economic growth. (Note that, while we have discussed national governments, for comparative purposes, this book also looks at two supra-national organizations: the European Union and the World Bank). Tony Meggs, Chief Executive of the UK's Infrastructure & Projects Authority (now NISTA, overseeing UK major government projects) wrote:

The vast majority of government policies are delivered through the implementation of a project or programme of some description. These projects and programmes span a wide range…[but] have one thing in common: if the projects are not successfully implemented, then the policy objectives are not delivered (Meggs 2018).

We thus need to be able to draw a clear line from government policies to projects undertaken to enact or support those polices and thence to the governance frameworks designed to ensure that those projects do indeed satisfy that requirement. While this is an easy statement to make, in some major projects and particularly in the biggest, 'megaprojects', the development of policies and governing those frameworks is a 'wicked' environment characterized by complexity, uncertainty, and conflict (Esposito and Terlizzi 2023) so that the development of a project and its governance will be a non-linear and conflicted process.

The sweeping definition of 'public projects', encompassing all major undertakings to deliver public policy, means that there is considerable variation in the type of project, and it is not self-evident that the same approach to governance is appropriate to all types of projects. The chapters in this book will describe the coverage of their country's governance scheme, which may not be applicable to all types of projects (sometimes termed programmes – we do not distinguish between these for this book). But to give a flavour of the types of project that can be covered, the UK Government's Major Projects Portfolio is categorized into four types of project (IPA 2023):

- Infrastructure and construction projects improving and maintaining the UK's infrastructure (energy, environment, transport, sewage, etc.);
- Transformation and service delivery projects, changing ways of working and harnessing new technology to improve public services;
- Military capability; and
- ICT (Information and Communication Technology) projects enabling the transition from legacy systems to new digital solutions.

The project governance framework put in place by a country's government might be applied to all projects or perhaps only projects above a certain 'size' (however that might be defined); on the other hand, it might be that the very largest projects, the so-called 'mega-projects', might be treated as special cases: they often have a higher public profile, and evidence suggests that for such projects, the governance framework should not be static but rather 'governance regimes must adapt to the specific project and context, deal with emergent complexity, and change as the project development process unfolds' (Miller and Hobbs 2005).

2.4 Project Life Cycle

The idea of a 'project' as an instrument to achieve an aim, along with how to carry out that project, is long-standing, with well-recognized ideas underpinning it. The life cycle of projects, shown in many project management textbooks, also has a general pattern that is common to most projects. The beginning of a project, often called the 'front end' of a project, is generally a time of definition, planning, and setting up before the project is executed in earnest. Following the front end, there is an execution phase where the work of the project is carried out, followed often by a phase of hand-over or passing the project to the permanent organization or to 'business-as-usual'. Good practice also suggests that later there should be an assessment phase to see how well the project was carried out, whether it was effective, and whether lessons can be drawn.

It has been well established that the front-end period is fundamental for the outworking, and in particular the success or failure, of the resulting project so that it will feature significantly in this book. There are various handbooks and papers on which activities in general should be carried out in the front end of a project (public or private); the following are summarized in Williams et al. (2019):

- The project initiative to start considering some sort of project: preliminaries (an initial business case or project proposal, and project selection and go/no-go decisions)
- Consideration of the project purpose: strategic alignment with the organization, success criteria, benefits/needs analysis, and stakeholder management
- Concept analysis and analysis of alternatives (consideration of concept, uncertainty, complexity, and evaluation)
- Assessment (defining and estimating, risk analysis, environment/sustainability, technology, and project delivery system)
- Setting up project execution (project finance, governance for the execution phase, and contracts/procurement).

When considering public sector projects, in principle, these follow the same life cycle as a generic project. But in the public sector, there are a number of specific considerations as the life cycle proceeds (see Williams (2022), although, to some extent, these might apply to some private sector projects also):

- The starting point will be some sort of government objective or policy intent. This might be specific and well-defined but can be vaguer and more generic, and the specification of this might be part of the planning process.
- The process of moving from policy intent through the front end, execution, and passing into 'business-as-usual' will often be governed by a formal framework. It is the variety of these in different countries and supra-national organizations, which is the topic of

this book. Tony Meggs, in the same blog as earlier, talks about the attempts to find 'a seamless flow and inter-connectivity between policy conception, policy development, and policy delivery' (Meggs 2018).

- For a public project, there will be a wide range of stakeholders, perhaps including politicians, members of the public, pressure groups, regulators, international bodies, local residents, and so on. These may have heterogeneous views about the objectives of the project – sometimes conflicting views, and some stakeholders might be opposed to the policy intent itself. The governance framework, in supporting the work of a national government, has to have a place to recognize these stakeholders. Indeed, it often needs to recognize the complexity of both internal and external stakeholders – Brunet et al. (2024) explore the complexity involved in collaborative governance in such projects, where multiple stakeholders collaborate and pool resources to further the project.
- On the other hand, Gil and Beckman (2025) talk about a 'governance trap', where 'managers, constrained by shareholder-centric investment choices that dismiss stakeholder collaboration costs as unnecessary expenditures, commit early on to tight budgets that exclude these costs to secure project approval', but since stakeholders can delay projects or require extra social and environmental concerns, so 'tight budgets trap managers in endless bargaining cycles with stakeholders until investors relax budget constraints' – that is, expected costs rise.
- Often, the execution phase of a project will be undertaken by a private concern such as a construction company; while this might be a public-spirited company, its objectives will be unlikely to be the same as the government's, so a contractual relationship will need to be developed to deliver the outcomes that the government desires. Governance of this relationship requires clear decisions about the type of relationship sought between the government and the contractor, be it more transactional or more collaborative. The prevailing wisdom in the public sector has varied between these two (Winch 2023).
- Public projects can often take many years, so the life cycle is likely to cross government budgetary cycles, maybe multiple different ministers, maybe even election cycles, so policy intent and financing might change during a project. (Indeed, the whole environment of a public project is likely to be turbulent.)
- Finally, a government needs to show that it is spending taxpayers' money wisely, so there is likely to be assessment of the efficiency and effectiveness of the project not only by the direct client but also by public bodies such as National Audits. A particular issue that has to be faced by public projects is assessing the value produced by a project, as 'value' to the government can be multi-faceted, ill-defined, and difficult to quantify, as we will discuss later in this chapter.

2.5 Governance of Single-Defined Public Projects

Before considering in the next section the issues around constructing a framework for governance of the portfolio of projects held by government, let us consider the issues around governance of a single project taken in isolation.

A starting indication of the role of a governance system is given by the Project Management Institute's PMBOK Guide, which assumes an organization has a system for delivering value and then says that the governance system works alongside this: 'Governance systems provide a framework with functions and processes that guide activities. A governance framework can include elements of oversight, control, value assessment, integration among components, and decision-making capabilities. Governance systems provide an integrated structure for evaluating changes, issues, and risks associated with the environment and any component in the value delivery system...' (Project Management Institute 2021, p. 12).

The governance system thus provides oversight of the activity generating *value* for the organization – or in the case of public projects, *value for the nation*. Within the public sector, this focus on the value of the project reveals a number of issues.

The project will first be aimed at satisfying some government objective, such as a policy intent, and this needs to be in alignment with the overall objectives of the government or the government department. This might be an unexpressed objective taken for granted, but it may be part of a government department's formal strategic plan. Many governments are 'gradually developing systems by which the outputs likely to accrue from projects are identified, quantified and linked to...strategic priorities', as discussed in Williams (2022, p. 16).

However, what is the 'value' of a public project? In the public sector, there is likely to be a high degree of heterogeneity in the types of value derived from a major project. Some will be simple monetary benefits that can be predicted and quantified: cost savings, efficiencies, income, etc. Others, however, may be ill-defined, even non-quantifiable, such as 'community cohesion' and 'feelings of safety'. 'When we seek to improve the lives of the citizens of a country, we are in territory that is subjective and contested. Identifying benefits is therefore a process that needs to engage a wide range of stakeholders' (Williams 2022, p. 17): stakeholders are likely to hold views that markedly differ from each other, maybe even be contradictory, as we noted in Section 2.3. The same reference gives an example of a road-tunnel project, beneath a World Heritage Site monument, and a panoply of officially recognized and measured benefit categories, including (for example), the health of the ecosystem, safety, use of non-motorized travel modes, access to the World Heritage Site, and so on. For further discussion, Brunet and Baba (2023) and Drouin and Genest (2023) describe approaches where stakeholders have different views of social acceptability, and specifically the issues of sustainability and the environment, evaluating these issues and gathering stakeholder views.

The net value of a project will be a combination of the positive benefits (and any disbenefits) of the project, and its costs, particularly financial costs (but possibly other classes of costs). Cost–benefit analysis can then be used to assess this (although this has weaknesses, including its normative foundation – see Volden 2019a). While, to some extent, some of these will be known in advance of the project, generally, the project has to go through its definition phase before a view can be taken of benefits and costs, and even then, they can only be predicted until the project has been executed, and even then, some will not yet be known. This brings us to issues around the estimation/prediction of both the positive benefits and costs.

Estimation of costs and benefits is carried out by individuals, subject to the various biases that affect all individuals. For Flyvbjerg et al. (2003, 2018), the root cause of project cost overruns is human biases, psychological, and political. He points to 'optimism bias' (the generally accepted tendency for humans to overestimate benefits and underestimate costs) and 'strategic misrepresentation' (deliberating biasing the estimates, perhaps to ensure a project gets funded that would not be if estimates were honest). If governments are to make rational decisions, Flyvbjerg says, the governance system needs to compensate for the former and identify the latter, and a government framework will need to indicate how this should be done. Flyvbjerg offers Reference Class Forecasting (RCF) (discussed in Flyvbjerg et al. 2018) as a partial solution, Williams (2022) notes that while other countries required optimism bias to be considered – often by applying an 'outside view', a review in some sense outside the project team – it seemes that the United Kingdom is alone in mandating the use of RCF.

The cause-and-effect relationship between bias and cost overruns has never been demonstrated (Flyvbjerg et al. 2018) is part of a heated exchange of academic papers with some authors, such as Love (specifically, Love and Ahiaga-Dagbui 2018), who call into question some of Flyvbjerg's methods and conclusions. These authors also point to Hirschman's fortuitous 'Hiding Hand' in projects, which highlights the creativity prompted in project teams by unexpected difficulties and obstacles. By considering both these theoretical insights and avoiding an 'either/or' stance, Ika et al. (2022) propose a balance between them accommodating both explanations as to why projects experience cost overruns (and benefit shortfalls), proposing a new principle they call the 'Fifth Hand', which allows for all possibilities in the development of projects and, for example, allows for various success/failure paradoxes. These arguments point to a connection with the critical issue of project governance as it relates to cost and benefit estimation and performance. In summary, Flyvbjerg and associates provide what some have called a 'governance paradigm' of cost overruns, 'which seeks to curb political and psychological factors and thus ensure accountability on the part of decision-makers' (see Love et al. 2019, p. 2589). In contrast, Love, Ika, and associates offer what they call a 'project management paradigm', 'which emphasises "best practices" for cost estimation and project execution as a way to counteract technical and economic causes of cost overruns' (again see Love et al. 2019, p. 2589).

2.6 Governance of Projects

This brings us to the crux of our subject: in order to provide governance for the portfolio of projects being undertaken by a government, what sort of framework or structure is needed? Klakegg et al. (2008) give their definition of a governance framework: 'an organized structure established as authoritative within the institution, comprising processes and rules established to ensure projects meet their purpose' (p. 27); they continue: 'the project meeting its purpose is a way of defining its success. It implies both delivering the relevant solution in an effective way and achieving a sustainable effect'.

The principles of what 'good' governance looks like, described earlier, are as relevant to project governance as to overall corporate governance. However, the nature of public projects, with their distinctive life cycle, wide variety of stakeholders, asymmetric information availability, asset specificity, and public interest, requires specific aspects to be considered in designing a governance framework. This section discusses this topic in general, leaving subsequent chapters to consider how these issues have been approached in specific countries.

Public projects have a long life cycle, and the governance framework can apply to all or just part of this life cycle. Public projects in general have a similar life-cycle, and where the governance regime applies, there will often be 'stage-gates' (Cooper et al. 2002), generally at decision points, where government can take a step back and decide whether or not the project should continue, or make an assessment on the progress and state of the project. If we look through the typical life cycle of projects, as described in Section 2.3:

- As discussed earlier, the genesis of any project is in a policy intent, or need, or desire on behalf of the government or a minister.
- The path from this requirement to a well-formed project, through identifying the various solutions that would satisfy the requirement or solve the problem, choosing one of these solutions to pursue, and crafting a well-formed project around this solution, can sometimes be a long and torturous journey. As also discussed earlier, it requires estimation of both the various benefits and the various costs that are likely to be involved, with the dangers of optimism bias and/or strategic misrepresentation due to personal or political motivations. It also requires some decisions on trade-offs, which are somewhat more subtle, as discussed in the study by Aubry and Floricel (2022), such as assessing when to move from exploring a variety of solutions to concentrating on a limited set, then one chosen solution; or the compromise between wide stakeholder consultation and concentrating on the interests of the commissioning government department. Furthermore, consideration of the project should not be in isolation but needs to be considered within the portfolio of projects held by the government, some of which might be related to the project under consideration, and all of which together are intended to move the government's purposes forward in a coherent manner.
- Even when the project is well-formed, the project front end (Williams et al. 2019) will require development of a project ready to execute, with decisions to make, such as the degree to which the project output is pre-defined before the project starts, as opposed to the desired outcomes being defined (again see Aubry and Floricel 2022).
- During the execution of the main project, governance is necessary to maintain the visibility of progress to the commissioning government, after which there is often some sort of handover so the project output becomes part of 'business-as-usual'. During the project, there might be regular reviews of progress and quality assurance, again better held by those with an 'outside view', reporting up to those accountable for the project outcome.
- But this is not the end of the project, as the need to demonstrate good stewardship of public funds and value-for-money means that there will often need to be 'ex post' assessments or reviews about how successful the project has been. This is discussed in the following section.

The point should be made that governance frameworks might not be static. The systems we will discuss within this book might need to encompass changes in project governance over time during the execution of a project. While this might not be relevant to medium- or maybe even large-sized projects, it is clear that the demands of the very biggest public projects ('mega-projects') and indeed the fundamental value propositions of such projects, change over time, and the governance structure needs to adapt appropriately (see Gil and Fu 2022). The dangers of misaligning the project governance and the changing project goals in mega-projects are highlighted in the study by Bourne et al. (2023).

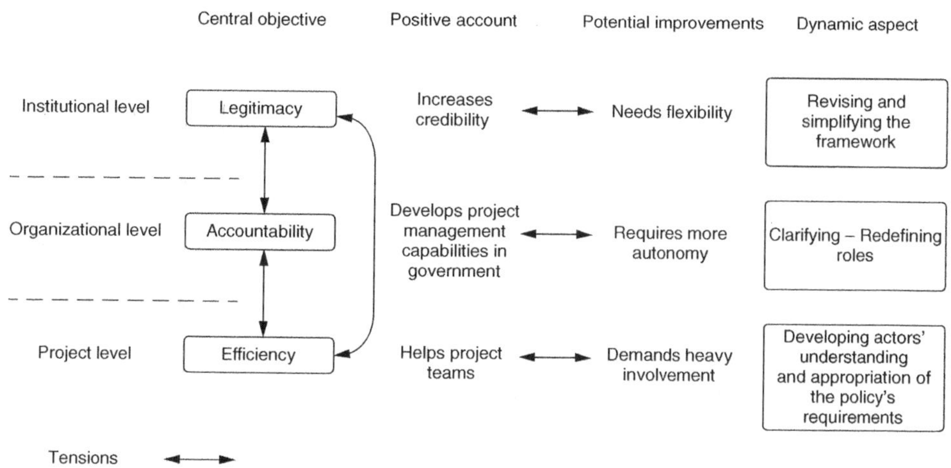

Figure 2.1 Multilevel model of the three objectives of a governance framework. *Source:* Brunet (2021) / with permission of Elsevier.

This is even more the case when major environmental changes occur during the life of the project (see the examples in Ika et al. 2024a). Furthermore, beyond the level of the single project, we will also be considering in this book how the governance frameworks themselves change over time in a specific country in reaction to policy, socio-economic, and environmental changes.

There can be a wide variety of governance frameworks, which can be more or less flexible. Still, there are clear benefits from developing some sort of framework, and these benefits can be seen as motivations for the developments described in the following chapters. Brunet and Aubry (2016) argue that in general 'a governance framework for major public projects can lead to', first, greater government efficiency (in the context of New Public Management and later New Public Governance, Osborne 2010); second, greater government legitimacy (e.g. by ensuring centralized political decision-making based on an analytical process); and third, greater government accountability (e.g. by having clear definition of roles and responsibilities and expected outcomes, and transparent evidence of the process). These three dimensions – efficiency, legitimacy, and accountability – come together in a useful framework shown in Figure 2 and Table 2 of Brunet (2021), with which we can compare frameworks discussed in this book (Figure 2.1).

2.7 Post-project Assessments

As we have said, the need to demonstrate good stewardship of public funds – that is, value-for-money – implies the need for 'ex post' reviews about how successful projects have been. As in Klakegg et al.'s 'delivering …in an effective way and achieving a sustainable effect' quoted earlier, 'success' implies both the achievement of the strategic intent of the project, and that the intended effect is sustainable, with minimal unforeseen disbenefits, but also that the project achieved its tactical 'iron-triangle' criteria by executing the project on-budget, on-time, on-quality; this latter aspect is sometimes referred to as 'project management success' rather than the wider 'project success'. All of these criteria can be arguable (e.g. How do you distinguish project effects from general economic effects? Does 'on-budget' mean the original budget at the start of the front end or the final agreed budget before execution, etc.), and the heterogeneous nature of stakeholders in any public project will often mean there are varying, contested views of what 'success' means. The meaning of 'project success' for public projects is discussed further in the study by Williams (2022). Still, this idea does not provide a comprehensive picture of a project's performance: the assessment of project performance, even retrospectively, is fraught with conceptual issues, and some of these are covered in the discussion in the editorial of Bosch-Rekveldt et al. (2023).

Such assessments are important to judge the worth of a project. But in the context of this book, these assessments over a range of projects are particularly important in considering whether the governance frameworks themselves are being useful. This is a difficult judgment to make in the absence of a natural experiment comparing projects within the framework to those out of the framework (beyond very simplistic assessments of cost and time overruns). Even the basic data on project performance to start such consideration may not exist, or be insufficiently complete. Then, as was noted earlier, the assessment of project performance is made difficult by many conceptual issues. Some work has been undertaken comparing projects before and after the introduction of various elements of a governance framework, such as the work of the Concept programme in Norway (see Chapter 3 of this book). The issue of the benefits of the frameworks and evidencing those benefits (and examining the disbenefits, such as the potential impact on risk appetite when civil servants are aware that their actions will be retrospectively reviewed) will be discussed later in the book.

2.8 Politics

We have discussed in this chapter only what a government mandates as a governance regime, or what official guidance might be. We have not discussed the questions of whether these requirements are actually followed in practice, and if not, how actual practices deviate from the espoused process and why people do not follow the process laid down. Indeed, all the countries involved have clear guidelines for governing projects. Williams et al. (2022) discuss a considerable range of paradoxes and issues that arise when the clear and logical plans laid down for project governance encounter actual practice, with its personalities, personal biases, and politics. Pinto and Davis (2023) show how what they call 'deviant behaviours' can not only be tolerated but also become normalized in practice.

One key area that clearly needs discussion is politics. Politics is often overlooked in discussions of governance frameworks, even though a key purpose of governance frameworks is to facilitate the implementation of politics and its corresponding policy decisions. However, when looking at the practical outturns of implementing such a framework, Samset and Volden's (2016) description of 'project paradoxes' includes the 'paradox of strategic alignment (p. 304) in most cases the internal logic of causalities and the probabilities of realization are erroneous', and they talk about decision-makers 'disregard[ing] analyses of costs and benefits', 'myopic decisions', and 'perverse incentives'. And there can be natural human resistance both to the introduction of a governance system and to its ongoing influence. So, chapters in this book will briefly consider the opportunities for untoward political influences affecting, disrupting or over-ruling the planned governance frameworks.

Experience shows that often major public projects lack well-defined benefits but rather are driven by political considerations – certainly, such considerations play a key role in how governance of projects plays out. Indeed, politics can sometimes play a more significant role than policies – one example is Stringer et al. (2014) looking at climate change, where they demonstrate good policies being pursued, but detrimental political interference getting in the way of the policy implementation.

Even without necessarily any perverse human behaviour being involved, the effect of external factors such as electoral cycles (and including strategic misrepresentation) can have an effect on the stability of policy implementation, as is shown empirically in the study by Catalão et al. (2022). Furthermore, any individual with decision-making authority is subject to human biases and emotions, one key perhaps being the well-known 'escalation of commitment' phenomenon highlighted in the work of Keil (e.g. Keil et al. 2000).

By contrast, long-term and international politics on climate change and other 'grand challenges' can be less vulnerable to the detrimental effects of electoral cycles but are subject to a nuanced array of influences by political actors, which complexifies governance and delivery of projects (see Ika et al. 2024b).

Of course, one aim of a national legal framework is to prevent these types of effects from tipping into corruption. An important, well-known phenomenon is 'pork-barrel politics' (Sharma 2017), which involves public funds being directed more towards specific local interests, irrespective of the overall public benefit. Akwei (2020) shows in one context that the reasons for failure of project delivery are 'political culture, political corruption, and poor planning and implementation, which are shaped by partisan politics in the Ghanaian context'. As an example of an attempt to thwart such effects, the Program Management Improvement and Accountability Act (USA Congress 2015) is now part of US law and 'mandates that every public institution define and implement procedures and standards for implementing projects and programs. That is, these cannot be based on direct interference from politicians' (Gasik 2023).

In summary, the governance system needs to align projects with the aims of the government, which by definition are political, with political priorities. However, there is clear scope for inappropriate political considerations to derail the use of a framework in practice, and we will need to consider the robustness of any framework against this.

2.9 Theories

There are three fundamental theories that are generally referred to in the literature when analysing project governance frameworks, and these can be useful in comparing the different emphases of the various frameworks.

Two of the theories, closely linked, are based on the idea of people as economic actors:

i) Transaction Cost Economics (TCE) was developed by Williamson in the 1970s and 1980s (e.g. Williamson 1979). It sees people as both boundedly rational and self-interested.

The theory focuses on the concept of transaction cost, which can be very significant in setting up a project (which may need to be planned, tendered, bid, and contracted). The governance approach used can be significant in determining the transaction cost incurred. According to Williamson (1979), there are a number of issues important in determining transaction costs: bounded rationality of the actors; information asymmetry between the actors; asset specificity, particularly relevant in one-off projects; frequency of exchanges; the atmosphere between the actors (i.e. of suspicion); and threat of opportunism (and additionally Small Number opportunism).

ii) The second idea is Agency Theory. In general, this theory considers the nature of the relationship between a principal and an agent when there is a contractual relationship between them. 'Because the unit of analysis is the contract governing the relationship between the principal and the agent, the focus of the theory is on determining the most efficient contract governing the principal-agent relationship given assumptions about people, …organizations …and information …' (Eisenhardt 1989 p. 58). The theory looks particularly at issues of goal incongruence between the parties, information asymmetry (i.e. it is difficult for the principal to verify what the agent is actually doing), and the problem of risk-sharing (when the parties have different attitudes towards risk). While Agency Theory is often used to describe the relationship between a client and contractor, it has been applied to the relationship between government/society and beneficiaries, for example, in Volden (2019b) on perverse incentives.

There is a third theory, chronologically and substantively complementary to the first two:

iii) The third idea is Stewardship Theory, which considers situations where managers and agents act as stewards aligning their motives with those of the principal. 'Economic approaches to governance such as agency theory tend to assume some form of homo-economicus, which depict subordinates as individualistic, opportunistic, and self-serving. Alternatively, sociological and psychological approaches to governance, such as stewardship theory, depict subordinates as collectivists, pro-organisational, and trustworthy', explains Davis et al. (1997, p. 20), who go on to 'explore the psychological and situational mechanisms that motivate stewards to behave pro-organizationally' rather than with self-interest.

We can see that there are different emphases in these theories: Transaction Cost Economics emphasizes efficiency, and Agency Theory emphasizes control; both are based on the ideas of the economic human actor; Stewardship Theory, using wider psychological ideas, emphasizes autonomy. This leads Joslin and Müller (2016), adapting Müller and Lecoeuvre (2014), to draw up a two-dimensional categorization of the underlying ideas behind governance structures: 'a) a continuum of the extent of shareholder versus stakeholder orientation …b) a continuum on the level of behaviour versus outcome control …as exercised by the project's parent organisation. …Agency theory is hereby seen as a proxy in explaining behaviour in more shareholder-oriented governance structures, where contracts and process control structures are used to manage the self-serving behaviour of managers for the maximization of shareholder wealth. Contrarily, stewardship theory is taken as a lens explaining behaviour in more stakeholder-oriented governance structures, where trust and controlling by outcomes/results serve as a mechanism to govern towards the achievement of organizational goals by balancing the requirements of a diverse set of stakeholders'. This can be used as a useful categorization of governance approaches.

While these are the fundamental theories that are generally referred to in governance literature, other theories are relevant to the matters in this book. In particular, we are discussing the introduction of governance frameworks and the responses and actions of the various actors within the governance system. One theoretical approach that is particularly relevant, therefore, is the projects-in-practice perspective (see Geraldi and Söderlund 2018). For the aspects in which this book is interested, this considers how actors enact the frameworks and apply them in day-to-day activities; a discussion of governance from this perspective is described in the study by Brunet (2019); for a more substantial discussion, see also Song et al. (2022).

2.10 Conclusion

These considerations have introduced us to some ideas that will enable us to consider and compare governance frameworks for public projects in different countries. Of course, these can only be a snapshot in time, as these frameworks are being developed in different ways at different rates in the various countries. Klakegg et al. (2016) took a look at that point in time at the governance regimes in three countries. They concluded 'that efforts to improve major projects are giving rewards: Even if complex public projects, the "trolls", become more challenging, efforts to "tame" them are improving' (p. 613). The following chapters of this book will review different frameworks and start to discuss their actuality in practice.

References

Akwei, C., Damoah, I.S., and Amankwah-Amoah, J. (2020). The effects of politics on the implementation of government programs/projects: insights from a developing economy. *Politics & Policy* 48 (6): 1161–1201.

Aubry, M. and Floricel, S. (2022). Undertaking the Project Front-End. Chapter 4. In: *The Front-end of Large Public Projects: Paradoxes and Ways Ahead* (ed. T.M. Williams, K. Samset, and G.H. Volden). Abingdon: Routledge/Taylor & Francis.

Bemelmans-Videc, M.-L., Rist, R.C., and Vedung, E. (1998). *Carrots, Sticks and Sermons. Policy Instruments and Their Evaluation*. New Brunswick, NJ: Transaction Publishers.

Bevir, M. (2012). *Governance: A Very Short Introduction*. Oxford: Oxford University Press.

Bosch-Rekveldt, M., Bourne, M., Forster, R. et al. (2023). Performance measurement in project management. *International Journal of Project Management* 41 (7): 102518.

Bourne, M., Bosch-Rekveldt, M., and Pesämaa, O. (2023). Moving goals and governance in megaprojects. *International Journal of Project Management* 41 (5): 102486.

Brunet, M. (2019). Governance-as-practice for major public infrastructure projects: a case of multilevel project governing. *International Journal of Project Management* 37 (2): 283–297. https://doi.org/10.1016/j.ijproman.2018.02.007.

Brunet, M. (2021). Making sense of a governance framework for megaprojects: the challenge of finding equilibrium. *International Journal of Project Management* 39 (4): 406–416.

Brunet, M. and Aubry, M. (2016). The three dimensions of a governance framework for major public projects. *International Journal of Project Management* 34: 1596–1607.

Brunet, M. and Baba, S. (2023). Social acceptability and governance for public infrastructure projects. In: *Research Handbook on the Governance of Projects* (ed. R. Müller, S. Sankaran, and N. Drouin). Cheltenham: Edward Elgar.

Brunet, M., Petit, M.-C., and Romero-Torres, A. (2024). Interorganizational design for collaborative governance in co-owned major projects: an engaged scholarship approach. *Project Management Journal* 55 (5): 580–598.

Cadbury, S.A. (2000). The corporate governance agenda. *Corporate Governance: An International Review* 8 (1): 7–15.

Catalão, F.P., Cruz, C.O., and Sarmento, J.M. (2022). Public management and cost overruns in public projects. *International Public Management Journal* 25 (5): 677–703.

Clegg, S. (2019). Governmentality. *Project Management Journal* 50 (3): 266–270.

Clegg, S.R., Pitsis, T.S., Rura-Polley, T., and Marosszeky, M. (2002). Governmentality matters: designing an alliance culture of interorganizational collaboration for managing projects. *Organization Studies* 23 (3): 317–337.

Cooper, R.G., Edgett, S.J., and Kleinschmidt, E.J. (2002). Optimizing the stage-gate process: what best-practice companies do – Parts 1 and 2. *Research Technology* 45: 21–27. and 43–49.

Davis, J.H., Schoorman, F.D., and Donaldson, L. (1997). Toward a stewardship theory of management. *Academy of Management Review* 22 (1): 20–47.

Drouin, N. and Genest, Y. (2023). Environmental, social, and governance criteria and their relationship with governance of major infrastructure projects. In: *Research Handbook on the Governance of Projects* (ed. R. Müller, S. Sankaran, and N. Drouin). Cheltenham: Edward Elgar.

Eisenhardt, K.M. (1989). Agency theory: an assessment and review. *Academy of Management Review* 14 (1): 57–74.

Esposito, G. and Terlizzi, A. (2023). Governing wickedness in megaprojects: discursive and institutional perspectives. *Policy and Society* 42 (2): 131–147.

Flyvbjerg, B., Bruzelius, N., and Rothengatter, W. (2003). *Megaprojects and Risk: An Anatomy of Ambition*. Cambridge: Cambridge University Press.

Flyvbjerg, B., Ansar, A., Budzier, A. et al. (2018). Five things you should know about cost overrun. *Transportation Research Part A: Policy and Practice* 118: 174–190.

Foucault, M. (2007). *Security, Territory, Population: Lectures at the Collège de France, 1977–78*. Springer.

Gasik, S. (2023). Some insights on the future of project management in public administration. *Revista de Gestão e Projetos* 14 (3): 27–39.

Geraldi, J. and Söderlund, J. (2018). Project studies: what it is, where it is going. *International Journal of Project Management* 36 (1): 55–70.

Gil, N. and Beckman, S. (2025). Escaping the governance trap: insights from new infrastructure development "megaprojects". *California Management Review* 67: 5–20.

Gil, N. and Fu, Y. (2022). Megaproject performance, value creation, and value distribution: an organizational governance perspective. *Academy of Management Discoveries* 8 (2): 224–251.

Ika, L.A., Love, P.E.D., and Pinto, J.K. (2022). Moving beyond the planning fallacy: the emergence of a new principle of project behavior. *IEEE Transactions on Engineering Management* 69 (6): 3310–3325.

Ika, L.A., Locatelli, G., and Drouin, N. (2024a). Policy-driven projects: empowering the world to confront grand challenges. *European Management Journal* 42 (6): 835–842.

Ika, L., Meredith, J., and Zwikael, O. (2024b). Project governance: the impact of environmental changes on governance adaptations in large-scale projects. *International Journal of Managing Projects in Business* 17 (4/5): 829–854.

IPA (2023). IPA Annual Report 2022–23. https://www.gov.uk/government/publications/infrastructure-and-projects-authority-annual-report-2022-23 (accessed 07 July 2024).

Joslin, R. and Müller, R. (2016). The relationship between project governance and project success. *International Journal of Project Management* 34 (4): 613–626.

Keil, M., Tan, B.C., Wei, K.K. et al. (2000). A cross-cultural study on escalation of commitment behavior in software projects. *MIS Quarterly* 24: 299–325.

Klakegg, O.J., Williams, T., Magnussen, O.M., and Glasspool, H. (2008). Governance frameworks for public project development and estimation. *Project Management Journal* 39: S27–S42.

Klakegg, O.J., Williams, T., and Shiferaw, A.T. (2016). Taming the 'trolls': major public projects in the making. *International Journal of Project Management* 34 (2): 282–296.

Love, P.E. and Ahiaga-Dagbui, D.D. (2018). Debunking fake news in a post-truth era: the plausible untruths of cost underestimation in transport infrastructure projects. *Transportation Research Part A: Policy and Practice* 113: 357–368.

Love, P.E., Ika, L.A., and Sing, M.C. (2019). Does the planning fallacy prevail in social infrastructure projects? Empirical evidence and competing explanations. *IEEE Transactions on Engineering Management* 69 (6): 2588–2602.

Meggs (2018). Civil service Quarterly blog "Crossing the 'Valley of Death' – how we can bridge the gap between policy creation and delivery". https://quarterly.blog.gov.uk/2018/03/28/crossing-the-valley-of-death-how-we-can-bridge-the-gap-between-policy-creation-and-delivery/ (accessed 07 July 2024).

Miller, R. and Hobbs, B. (2005). Governance regimes for large complex projects. *Project Management Journal* 36 (3): 42–50.

Müller, R. and Lecoeuvre, L. (2014). Operationalizing governance categories of projects. *International Journal of Project Management* 32 (8): 1346–1357.

Müller, R., Sankaran, S., and Drouin, N. (2023). *Research Handbook on the Governance of Projects*. Cheltenham: Edward Elgar.

Osborne, S.P. (ed.) (2010). *The New Public Governance? Emerging Perspectives on the Theory and Practice of Public Governance*. New York, NY: Routledge.

Osborne, S.P., Nasi, G., and Powell, M. (2021). Beyond co-production: value creation and public services. *Public Administration* 99 (4): 641–657.

Pinto, J.K. and Davis, K. (2023). Normalization of deviance in projects: its causes and implications for effective governance. In: *Research Handbook on the Governance of Projects* (ed. R. Müller, S. Sankaran, and N. Drouin). Cheltenham: Edward Elgar.

Project Management Institute (2021). *A Guide to the Project Management Body of Knowledge (PMBOK)*, 7e. Newtown Square, PA: Project Management Institute.

Samset, K.F. and Volden, G.H. (2013). Investing for Impact. Lessons with the Norwegian State Project Model and the first investment projects that have been subjected to external quality assurance. Concept rapport number 36, NTNU, Trondheim, Norway.

Samset, K.F. and Volden, G.H. (2016). Front-end definition of projects: ten paradoxes and some reflections regarding project management and project governance. *International Journal of Project Management* 34 (2): 297–313.

Samset, K.F., Volden, G.H., Olsson, N., and Kvalheim, E.V. (2016). Governance Schemes for major public investment projects: a comparative study of principles and practices in six countries. Concept Report no. 47, Trondheim: Ex Ante Academic Publishers.

Sharma, C.K. (2017). A situational theory of pork-barrel politics: the shifting logic of discretionary allocations in India. *India Review* 16 (1): 14–41.

Song, J., Song, L., Liu, H. et al. (2022). Rethinking project governance: incorporating contextual and practice-based views. *International Journal of Project Management* 40 (4): 332–346.

Stringer, L.C., Dougill, A.J., Dyer, J.C. et al. (2014). Advancing climate compatible development: lessons from southern Africa. *Regional Environmental Change* 14 (2): 713–725.

Tricker, R.I. (2015). *Corporate Governance: Principles, Policies, and Practices*. Oxford University Press.

USA Congress (2015). *Program Management Improvement and Accountability Act, PMIAA*. US Congress: Washington, DC.

Volden, G.H. (2019a). Assessing public projects' value for money: an empirical study of the usefulness of cost–benefit analyses in decision-making. *International Journal of Project Management* 37 (4): 549–564.

Volden, G.H. (2019b). Public funding, perverse incentives, and counterproductive outcomes. *International Journal of Managing Projects in Business* 12 (2): 466–486.

Williams, T.M. (2022). Project success. Chapter 2. In: *The Front-end of Large Public Projects: Paradoxes and Ways Ahead* (ed. T.M. Williams, K. Samset, and G.H. Volden). Abingdon: Routledge/Taylor & Francis.

Williams, T.M., Vo, H., Samset, K., and Edkins, A. (2019). The front-end of projects: a systematic literature review and structuring. *Production Planning and Control* 30 (14): 1137–1169.

Williams, T.M., Samset, K., and Volden, G.H. (2022). *The Front-end of Large Public Projects: Paradoxes and Ways Ahead*. Abingdon: Routledge/Taylor & Francis.

Williamson, O.E. (1979). Transaction-cost economics: the governance of contractual relations. *The Journal of Law and Economics* 22 (2): 233–261.

Winch, G.M. (2023). Governance across the commercial interface on complex engineering projects. In: *Research Handbook on the Governance of Projects* (ed. R. Müller, S. Sankaran, and N. Drouin). Cheltenham: Edward Elgar.

Part II

Case Studies of Project Governance Models

3

Norway

Gro Holst Volden and Morten Welde

Concept Research Programme, Norwegian University of Science and Technology (NTNU), Trondheim, Norway

Norway's approach to public investment management is shaped by a unique combination of economic strength, trust in institutions, and cultural values such as fairness and transparency. As a small but wealthy country with a robust welfare system and substantial oil revenues, Norway faces both exciting opportunities and complex challenges in managing large government projects. The development of the State Project Model – Norway's approach to planning and ensuring quality in large public investments – has enabled responsible government agencies and ministries to promote good governance, thereby increasing the success rate of the projects they are responsible for. This chapter provides a closer examination of the model's background, structure, and impact, offering insights into how Norway has incorporated best practices into its public sector project management.

3.1 Context and Policy Basis

Norway is a small (population 5.6 million) and wealthy country. It is ranked among the top 10 countries in the world, as measured by gross domestic product (GDP) at purchasing power parity per capita (World Bank 2024). It is politically stable and has low levels of corruption (Transparency International 2023). The country has experienced strong and sustained economic growth since oil and gas were discovered on its continental shelf in the late 1960s. Today, the net cash flow from the petroleum industry is estimated to be approximately NOK 670 billion per year, accounting for around 20% of the country's GDP (Norwegian Petroleum 2024). Norway is considered the Organisation for Economic Co-operation and Development (OECD)'s leading country in public spending, partly due to its extensive welfare system coverage and ambitious regional development goals. Norway's financial position is strong, and its oil revenues are accumulated in a sovereign wealth fund. However, challenges may lie ahead due to the gradual depletion of oil resources and the fiscal impact of an ageing population, which may require potential cuts in public spending (OECD 2024).

The Norwegian government consists of three levels: the central government, 15 counties, and 357 municipalities. Most taxes are collected by the central government,

Governance Frameworks for Major Public Projects: International Practices and Experiences, First Edition.
Edited by Vedran Zerjav, Morten Welde and Gro Holst Volden.

which then redistributes the revenues to the regional and local levels. This means that government-funded projects are crucial for achieving a desired change locally. In Norway, the local level has substantial influence over the design and implementation of government-funded projects, particularly because plans must be approved locally under the Planning and Building Act.

Norway's gross public investment is notably high in international comparison, particularly in transport infrastructure. Despite this significant spending, Norway's transport network does not rank well according to some indicators (OECD 2024). The country's low population density (15 per km^2), long distances, and challenging topography mean that few public infrastructure investments are considered value for money in cost–benefit analysis terms. For instance, the government's latest strategic transport plan for 2025–2036 proposed new projects with a combined net benefit–cost ratio of −0.37. Poor economic efficiency of investments can also be explained by inefficient planning practices, robust state finances, and regional policy objectives that prioritize a fair distribution of government funds over promoting economic efficiency (Sager 2016).

Norway's level of trust in institutions and other people is one of the highest in the world. Of note, 68% of Norwegians report trusting people they meet for the first time, and 64% of the population expresses confidence in the national government (OECD 2022; The Policy Institute 2023; OECD 2024).

The Norwegian work culture is characterized by egalitarianism and independence, meaning people generally prefer not to be told how to perform their jobs. The work culture prioritizes consensus-building and cooperation rather than individual achievement or competition (European Union 2023). The Scandinavian tradition differs from the Anglo-American tradition in that there is less emphasis on individual accountability. Norwegians tend to blame the system rather than the individual when things go wrong. The low risk of personal consequences encourages trust and transparency between individuals and between practice and academia. The culture of openness is further reinforced by the Freedom of Information Act, which states that all case documents of public authorities are publicly accessible unless specifically exempted.

3.2 The State Project Model

The State Project Model sets requirements for methodology and quality when appraising and planning large government investment projects. It is based on a 16-page government circular (Ministry of Finance 2025). It can be seen as a specialization of the more general requirements for appraisal and evaluation, as well as prudent economic management, as outlined in the Instructions for Official Studies and Reports (Utredningsinstruksen) and The Regulations on Financial Management in Central Government (Økonomiregelverket).

In addition to the formal requirements of the State Project Model, each ministry and agency may have its own (more detailed) project model, which includes regulations related to statutory requirements such as the Planning and Building Act (Volden and Andersen 2018). Similarly, municipalities and counties have their own governance frameworks for locally funded projects.

3.2.1 A Brief Historical Overview

The Norwegian model was introduced against a backdrop of numerous negative experiences with public investment projects in the 1990s, including cost overruns, delays, and benefits that did not materialize. In 1997, an interdepartmental committee was established to analyse a selection of projects to clarify what had gone wrong. The committee, which the Ministry of Finance chaired, concluded that the basis for decision-making was inadequate and that poor front-end planning was the main reason why projects failed. Too many projects were initiated by agencies and local stakeholders with little consideration for strategic alignment and value for money. Projects were appraised and planned for years, only to be presented almost as a *fait accompli* to the government, which was reluctant to reject projects with much political and popular momentum.

The Ministry of Finance introduced the first version of the framework in 2000. The primary requirement was that the cost estimates undergo external quality assurance (QA, later referred to as QA2) before a project could be submitted to the parliament for approval and funding. The intention was to root out over-optimism through an outside view provided by external experts (consultants) in project management and cost estimation from the private sector. The framework was inspired by similar schemes in the oil and gas industry.

In 2005, the model was extended to include an earlier quality assurance (QA1) before the Government decided whether to start the pre-project phase. The purpose of the wider model was also to ensure that the right projects were selected for further development and to improve value for money from public investments. In other words, the framework had shifted from a narrow focus on 'doing it right' to a broader perspective on 'doing the right thing'.

For many years, the framework was referred to as 'the QA scheme', consisting mainly of QAs before the two decision gates. Over time, however, there has been an evolution, with increasing attention from the Ministry of Finance and others on the exchange of experience, the development of guidelines, and other efforts to help actors succeed and harmonize practice. In 2019, the framework was formalized through a government circular, and at the same time, its name was changed from the QA scheme to the State Project Model. This implied a stronger formal requirement from the government.

3.2.2 The Stage-Gate Model with Phases and Decision Points

Figure 3.1 presents the current State Project Model. The model applies in principle to all major government-funded projects across sectors. The model applies to the front-end phase only, i.e. from the project idea is conceived until the funding decision is made, and not to project implementation or the operational phase that follows.

The model requires that projects be planned and carried out according to the phases shown in the figure: idea phase, conceptual phase, pre-project phase, construction phase, and operational phase.

3.2.3 Idea Phase

The idea phase involves the earliest stages of exploration to determine whether a potential problem exists or will arise that may or may not warrant an investment project. This phase

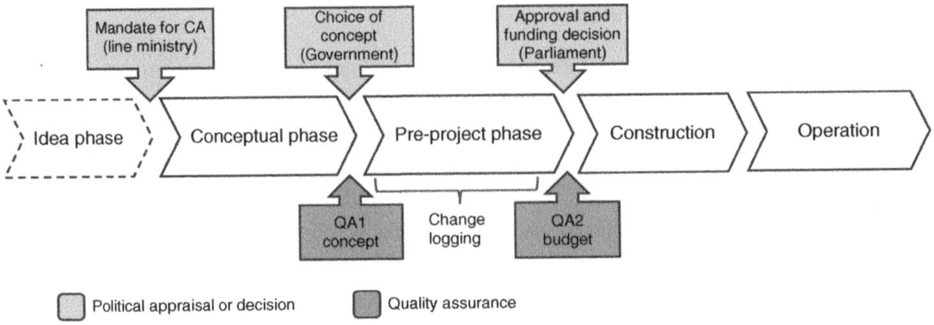

Figure 3.1 The Norwegian State Project Model.

has traditionally been (and still is) considered part of the 'informal' front-end phase, with no requirements for specific analyses or processes. However, the conceptual phase cannot start without a formal decision by the line ministry and a mandate being issued.

3.2.4 Conceptual Phase

The problem is investigated more thoroughly in the conceptual phase through a Conceptual Appraisal (CA). The ministry is responsible for the CA report, but often delegates it to a subordinate agency. The CA must be structured in the following chapters:

1) Description of the problem
2) Needs analysis
3) Strategic goals
4) Framework conditions
5) Exploration of opportunity space – identify at least three conceptually different alternatives, including the do-nothing alternative
6) Cost–benefit analysis of the shortlisted options
7) Recommendations for the preliminary project phase

Independent experts review the CA report through QA1, normally after the full CA report is published. The line ministry may opt for a review in two stages, where steps 1–6 are reviewed first and step 7 is reviewed in the second stage. This may be relevant when the concept choice is a very high-level decision, and a more detailed assessment of the selected concept is needed before the pre-project phase can start.

Finally, the government selects the concept and decides whether to move to the pre-project phase. The government's concept choice is not a formal approval or funding decision, but in practice, a green light for further planning may increase stakeholder expectations and political commitment, making final project approval more likely.

3.2.5 Pre-project Phase and Investment Decision

During the pre-project phase, the agency typically develops a full business case (FBC) that describes the project in greater detail, including more detailed estimates of costs and benefits.

Since 2019, line ministries have been required to set a target cost for the pre-project stage, and agencies must keep a log of changes for subsequent planning. This requirement came as a response to widespread scope creep and cost increases during the early phases of project development. It helps the owner (the line ministry) gain a better understanding of the project's content, scope, and consequences.

Based on a complete cost estimate, the FBC includes a recommended total budget (including contingency) and a lower target cost, or project budget, for the agency. The budget will typically be set at or near the P85 estimate, while the target cost will be slightly lower, at P50. A stochastic cost estimate at PX means that there is an X% probability that the actual cost will be at or below the estimated amount. *The business case* includes an updated cost–benefit analysis and often a benefits realization plan.

The FBC is then reviewed through QA2 before the line ministry can submit the project to parliament for final and formal approval. Formal project approval is not until after QA2. The parliament's funding decision is at a relatively mature project stage, typically before detailed design and prior to tender. Welde and Klakegg (2024) argued that the maturity of Norwegian cost estimates at the funding decision is equivalent to AACE Class 3 estimates and to the FBC in the UK stage-gate model.

3.2.6 Construction and Operation

According to the State Project Model, there are no compulsory follow-up points in the execution phase or after completion. This is based on the assumption that once the project is well justified and thoroughly planned, individual agencies have sound procedures for safeguarding the implementation and benefits realization.

3.2.7 Projects That Must Adhere to the Model

The State Project Model is mandatory for all government-funded projects with a presumed budget above NOK 1 billion. Line ministries are also free to apply the model to smaller projects if they consider it useful.

There are, however, some exceptions. The threshold value for digitalization projects is NOK 300 million. Projects justified on a commercial basis (e.g. offshore oil and gas investments under the auspices of the State's Direct Financial Interest) are exempted, and so are projects in state-owned enterprises whose goal is the most efficient attainment of public policy goals. Larsen et al. (2021) provide an overview of project governance frameworks in Norway's state-owned enterprises.

The model primarily encompasses transport projects (excluding aviation), defence projects, central government construction projects, and major information and communication technology (ICT) projects. The number of projects that undergo the model may vary, but on average, about 20 projects enter the State Project Model each year.

The line ministries can seek exemption from the Ministry of Finance from full CA/QA1 in cases where alternative conceptual solutions do not exist, such as if the remaining section of a new highway needs to be built and no other transport forms are available.

3.2.8 Parties and Roles

The organization of the model within the central government is illustrated in Figure 3.2.

- The **Ministry of Finance** is responsible for the government circular and the framework agreements with quality assurers.
- The **line ministries** and **government agencies** produce the input to the QA reviews and are also responsible for following up on the resulting recommendations. The line ministries are the true owners of any new investment initiative; however, in practice, they often delegate appraisal, planning, and implementation to subordinate agencies.
- Each QA report is commissioned jointly by the Ministry of Finance and the line ministry.
- The **quality assurers** review the documentation, undertake their independent analyses when needed, and, finally, give their recommendations.
- Based on the planning documents from the agency/ministry and the QA report, a proposal is submitted to the **political level** (i.e. the Government after QA1 and parliament after QA2).
- In parallel with the governance framework, a **trailing research programme**[1] was established to accumulate information about the projects and their performance over time and to develop knowledge about the effect of QA and other measures taken during the front-end phase. The research programme is independent of the governance model but is funded by the Ministry of Finance and plays a key role in data collection, ex post evaluation, and dissemination of good practice. The government circular states that QA reports and other project documentation must be delivered to the Concept Research Programme.
- The QA reports are, with some exceptions (notably defence projects), public and are published online.

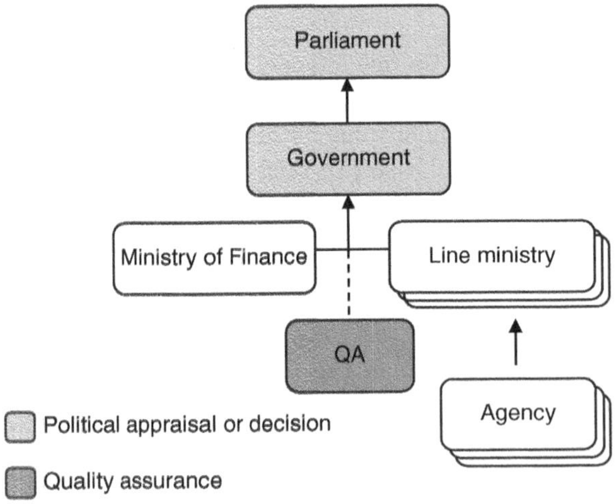

Figure 3.2 Investment project governance.

1 Concept Research Programme: https://www.ntnu.edu/concept.

3.2.9 Selected Topics

3.2.9.1 How Projects Are Initiated

Traditionally, many government-funded projects were local initiatives based on real or perceived needs, such as poor accessibility to national and international markets due to inadequate road standards or worn-down university facilities, museums, and other institutions. The State Project Model was introduced to manage initiatives at the government level early on, in order to avoid potential escalation of commitment later. The strategic perspective has been further strengthened through long-term investment plans in sectors such as transport and defence. However, despite these attempts, political decisions, especially in the transport sector, are deeply influenced by local electoral support. Changing Norwegian governments have tended to favour investments in regions where they enjoy local voter support.

Until recently, CAs were conducted to signal political ambition, even if project funding was not forthcoming. Volden et al. (2024) demonstrated that this had led to potential projects piling up and effectively ending up in limbo, because projects are not 'killed' but just 'paused' indefinitely. As a consequence, ministries now must treat the initiation of an appraisal as a formal decision. The mandate for appraisal outlines the direction, the problem for which an appraisal is needed, and the potential goals a project should achieve.

3.2.9.2 Option Appraisal

The Norwegian model requires the assessment of at least three conceptually different options, including the do-nothing alternative. For example, if the problem is a congested road, alternative solutions may include increasing road capacity, improving public transport, and even road pricing. Rather than starting with a specific solution, the CA should explore alternative options and assess their feasibility to ensure good use of public funds. The do-nothing alternative, which typically involves maintaining the current situation, should always remain a viable option. However, as the do-nothing can only include committed investments, a deterioration compared with the current situation must be expected.

3.2.9.3 Cost Estimation and Budget Management

Cost performance is a crucial element of the Norwegian State Project Model. Government agencies spend significant resources on estimating project costs before QA2 and parliamentary approval. At the CA/QA1 stage, the same methods are applied, but a higher-level (big picture) approach is taken.

Various forms of stochastic cost estimation have been common in some sectors in Norway since the 1990s and later formalized as part of the QA2 scheme. Estimating costs and their uncertainties is based on a top-down approach using triple estimates of known elements. The uncertainty is estimated through a group process and Monte Carlo simulation. The estimation process produces a range of outcomes with assigned probabilities.

The total budget for projects is normally at the P85 level. It includes a contingency. The P50 acts as a target for the responsible agency, while the project manager may be allocated a smaller budget.

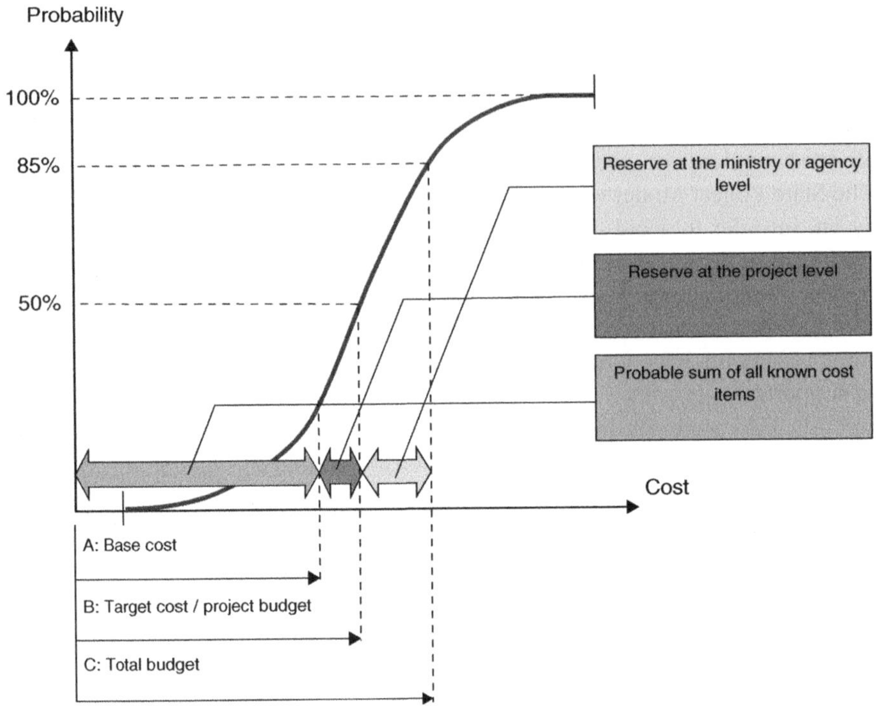

Figure 3.3 Stochastic cost estimation.

Figure 3.3 illustrates the outcome of a cost estimation process in Norway, showing a cumulative probability distribution of the investment cost.

The contingency, the difference between the P85 and the P50, is always allocated at an organizational level above the project. Generally, it is up to the line ministry as the project owner to decide whether to delegate it to the agency level. Investment projects within the transport and defence sectors are typically funded as part of a plan that includes several projects. The agency then receives the total budget (including contingency) but is still expected to manage its portfolio to ensure that all projects, viewed collectively, are completed within P50. The contingency is reserved at the ministry level in agencies that carry out fewer projects.

3.2.9.4 Quality Assurance of Decision Documents

External private sector experts with expertise in project management, cost estimation, economics, and related areas conduct the QAs. The current framework agreement includes seven constellations of consultants. The Ministry of Finance has issued six calls for tenders since 2000, and several of the consultancies have been part of the model since the start.

The appendices to the framework agreement specify the detailed requirements for what to assess as part of QA1 and QA2. The requirements are consistent with the government circular.

Each QA is relatively comprehensive. A QA2 report is typically 50–75 pages long plus appendices, and a QA1 report can be even longer. It involves reviewing the documents and an independent analysis of costs and benefits.

An evaluation in 2016 found that each QA1 took, on average, 290 days and cost NOK 2.45 million. The corresponding figures for QA2 were 195 days and 2.1 million NOK. The numbers demonstrate that QA in the Norwegian model is very resource-demanding. In recent years, efforts have been made to streamline the process.

3.2.9.5 Follow-Up and Ex Post Evaluation

The model applies only to the front-end phase. However, under the auspices of the Concept Research Programme, systematic ex post evaluation was introduced in 2012. Up to four projects are evaluated annually based on a standardized evaluation model that includes six criteria covering operational, tactical, and strategic success (Volden and Welde 2022).

As of 2025, over 45 projects had been evaluated. The Concept website provides a valuable database of lessons learned from these evaluations. An ongoing effort is made to involve affected parties, such as ministries and agencies, in the evaluation process, and the findings are shared and discussed with them afterwards. Additionally, Concept produces summaries of lessons learned and meta-evaluations based on multiple evaluations to support learning.

3.2.9.6 Guidance, Training, and Exchange of Experiences

The Ministry of Finance, as system owner, did two important things that have indirectly contributed to improving skills and encouraging learning. First, introducing the model incentivized ministries and agencies to improve the quality of their projects at entry. Second, a research programme was established to follow the framework and organize networking and dissemination activities.

The ministry itself has applied knowledge generated from research when updating the framework agreements every four years. With assistance from the Concept Research Programme, the ministry also arranges a biannual national conference attended by employees of all agencies and ministries involved in the framework. At the conference, various aspects of the model are discussed, including the need for guidance concerning different elements of the appraisals.

Initially, the Ministry of Finance was reluctant to issue instructions and guidelines. Gradually, this has changed. The Ministry is now launching a more comprehensive guide to CAs. There has also been discussion about the need to educate project managers and project owners; however, this is not part of the current framework.

3.2.9.7 Portfolio Management

The Norwegian model is focused on requirements applicable to individual projects only. There are no formal requirements for portfolio management as part of the scheme, neither on the agency level nor concerning the government's total portfolio of large projects. Some ministries and agencies with many large projects have started to assess and manage them as a portfolio.

3.3 Experience with the Model

The State Project Model is mandatory for all large projects in the sectors it covers, and only the Ministry of Finance can grant an exemption from the model.

Over the 25 years since the scheme was introduced, approximately 250 projects have undergone QA2, about 150 QA1 and 60 both QA1 and QA2. About half of the projects fall under the auspices of the Ministry of Transport, which has thus gained more experience with the framework than any other ministry.

3.3.1 Improved Practices

The model's starting point was the need to reduce the extent of cost overruns. Several studies have documented that this has largely been achieved. Welde and Klakegg (2024) documented that in a sample of 96 projects that had undergone QA2 and were now completed, the average cost *underrun* was 4%, and 73% of projects were completed within the total budget. The distribution around the P50 estimate was slightly skewed to the right. The median cost overrun was 5%, and almost 60% of the projects had final costs above the P50. The authors concluded that although there was room for improvement, the results were better than those reported in most international studies of cost performance.

The wider issue of strategic success can only be assessed through a comprehensive ex post evaluation. QA1 was introduced later, and there are still relatively few projects that have been through the full model and are in their operational phase. However, since the model's implementation and with gradual improvements over time, ministries and agencies have made considerable efforts to improve skills and practices. The old way of selecting the solution before an appraisal is no longer possible. Although sectoral responsibilities, such as those for road and rail, may lead to tunnel vision, there is an awareness that there are always different solutions to a problem.

Several studies from different countries have documented that appraisal of costs and benefits may be vulnerable to over-optimism or wilful deception. External QA has been an effective remedy for debiasing estimates and promoting good practice. As such, the Norwegian model has improved the quality of front-end management through a combination of improved internal practices (the inside view) and an outside view.

The framework has also improved the transparency of project planning and appraisal. Although the Freedom of Information Act states that most appraisal documents shall be publicly accessible, this may be a theoretical concept if the information is scattered in hundreds of documents buried in hard-to-access databases. The process for appraisal and QA is now standardized and common to projects in all sectors, which ensures comparability between projects. And since CAs and QA reports are publicly available online, as are ex post evaluations by Concept, this provides the media and other stakeholders with insight into the projects that would otherwise be very difficult to attain. Implicitly, it will contribute to accountability among civil servants, quality assurers, and decision-makers.

Improved practices in ministries and agencies have also had a disciplining effect on politicians, who now have less scope to sidestep the appraisal process and fast-track projects for approval. If a large project is to be implemented, it must be planned and scrutinized within the governance framework. This throws a bit of 'sand in the gears', which may curb the immediate enthusiasm that often surrounds project initiatives.

The model has also had a trickle-down effect, creating an awareness of the importance of the front end for project success, both within the public and private sectors. Over the past decade, several state-owned enterprises, as well as large municipalities, have also introduced similar frameworks to improve their practices.

Finally, research conducted by the Concept Research Programme is being used in teaching at universities and in-house training in ministries and agencies. Research results based on empirical data have been an invaluable source of knowledge on project performance, providing a more balanced and nuanced perspective than what is often reported in the media. It should also be noted that research from Concept has not only sparked debate in and given inspiration to other countries, such as Denmark, but also had an effect that extends across borders.

3.3.2 Remaining Challenges

When the project model was introduced, the focus was on reducing cost overruns. This has largely been achieved thanks to QA2, although the proportion of projects with a final cost exceeding budget is higher than desirable. However, there was less emphasis on controlling costs and scope changes before final budget approval. Several studies by the Concept Research Programme documented that costs have increased by an average of 40% between QA1 and QA2. This is akin to the Hawthorne effect, which states that individuals or groups modify an aspect of their behaviour in response to their awareness of being observed. In other words, cost performance during the implementation phase improved because it was monitored closely. Cost performance in the front end, however, improved much less because there was less emphasis on it. As we touched on earlier, the Ministry of Finance has also introduced measures to curb scope and cost creep between QA1 and QA2.

Although the requirement to explore conceptually different options has led to broader appraisals, we still see signs of path dependency and tunnel vision. CAs are typically delegated to an agency with specialized knowledge in certain areas and not others, which influences the solutions they choose to explore. Even if the Ministry of Finance's guidelines clearly state that a do-nothing alternative should always be included in the appraisal, Volden et al. (2024) documented that the overwhelming majority of appraisals recommend further planning of investment alternatives. Do-nothing or do-minimum options are rarely recommended by agencies or selected by politicians, even if they often provide better value for money than grand projects. This was a reason for formalizing the starting point of appraisal in 2019 (the mandate). In other words, to limit the number of candidate projects, we must control the number of appraisals.

With most focus on costs, benefits may have been overlooked in the early years of the model. Evaluations of the projects have shown that while some sectors follow up on the intended benefits after the project is approved, others rarely do so. Benefits management, as part of the programme and portfolio management in the agencies, is currently a priority for the actors in the model.

3.3.3 Widely Accepted and Under Continuous Improvement

When it was first introduced 25 years ago, the model was met with controversy. In line with the country's egalitarian values, civil servants did not want to be told how to plan and implement their projects. However, the attitude towards the framework has changed, and all involved parties widely accept the State Project Model.

The Ministry of Finance likely made a good decision by introducing only the QA elements initially, without providing detailed requirements for the ministries. The principle was "business as usual" – that is, the Ministry of Finance, as scheme owner, chose a buy-in approach instead of a coercive approach. Implicitly, it was assumed that the QAs would have a disciplining effect and that agencies would take action to improve their practices, which they did. Today, ministries and agencies have streamlined their processes, and the Ministry of Finance has issued more comprehensive guidelines.

The model fits well with requirements for traditional projects, as was the intention at its introduction. However, it may fit less well with agile digitalization projects and other public measures that are not covered. Some actors still make complaints about the model being time and resource-consuming and that the use of a threshold value is inflexible. The State Project Model, like other similar project governance models, should never be static, and there is still room for adjustments and improvements.

References

European Union (2023). Work culture in Norway. https://academy.europa.eu/mod/page/view.php?id=17995 (accessed 8 July 2024).

Larsen, A.S.A., Volden, G.H., and Andersen, B. (2021). Project governance in state-owned enterprises: the case of major public projects' governance arrangements and quality assurance schemes. *Administrative Sciences* 11: 66. https://doi.org/10.3390/admsci11030066.

Ministry of Finance (2025). *Rundskriv: Statens prosjektmodell - Krav til utredning, planlegging og kvalitetssikring av store investeringsprosjekter i staten* [Government circular. The State Project Model – Requirements for investigation, planning and quality assurance of large public investment project], R-108-25.

Norwegian Petroleum (2024). The Government's Revenues. https://www.norskpetroleum.no/en/economy/governments-revenues/ (accessed 8 July 2024).

OECD (2022). Trust in government. https://www.oecd.org/en/data/indicators/trust-in-government.html (accessed 8 July 2024).

OECD (2024). *OECD Economic Surveys: Norway 2024*. Paris: OECD Publishing. https://doi.org/10.1787/cb13475f-en.

Sager, T.Ø. (2016). Why don't cost-benefit results count for more? The case of Norwegian road investment priorities. *Urban Planning and Transport Research* 4: 101–121.

The Policy Institute (2023). *The State of Social Trust: How the UK Compares Internationally.* Kings College London. https://doi.org/10.18742/pub01-135.

Transparency International (2023). Corruption Perceptions Index. https://www.transparency.org/en/cpi/2023 (accessed 8 July 2024).

Volden, G.H. and Andersen, B. (2018). The hierarchy of public project governance frameworks. An empirical study of principles and practices in Norwegian ministries and agencies. *International Journal of Managing Projects in Business* 11: 174–197.

Volden, G.H. and Welde, M. (2022). Public project success? Measuring the nuances of success through ex post evaluation. *International Journal of Project Management* 40: 703–714.

Volden, G.H., Welde, M., Engebø, A., and Andersen, B.S. (2024). Do nothing, do minimum or do something? Why public project appraisals "always" recommend large projects. *International Journal of Managing Projects in Business* 17: 430–454.

Welde, M. and Klakegg, O.J. (2024). Avoiding cost overrun through stochastic cost estimation and external quality assurance. *IEEE Transactions on Engineering Management* 71: 1984–1997.

World Bank (2024). GDP per capita, PPP (current international $). GDP per capita, PPP (current international $) - Norway | Data (accessed 8 July 2024).

4

The United Kingdom

Richard Kirkham[1] and Nicola Young[2]

[1] *Department of Civil Engineering and Management, School of Engineering, The University of Manchester, Manchester, UK*
[2] *Thomas Ashton Institute, School of Engineering, The University of Manchester*

The United Kingdom is widely regarded as an early adopter and leading international exemplar of a public investment governance scheme that is designed to ensure scrutiny in the delivery of major public investments. The scheme is designed to focus on the Government Major Projects Portfolio (GMPP), presently containing 213 projects subject to a rigorous oversight process jointly administered by a dedicated body on behalf of His Majesty's Treasury (HMT) and the Cabinet Office (CO). All UK government departments, their contractors, and many other organizations use a similar/same process. Since 2025, the National Infrastructure and Service Transformation Authority (NISTA), formerly known as the Infrastructure and Projects Authority (IPA), has provided the expertise and central coordinating function necessary to enable the governance of the GMPP. This chapter outlines the background of the scheme and its main features and concludes with some experiences of its implementation.

4.1 Context and Policy Basis

The United Kingdom is a relatively prosperous economy, underpinned by technological innovation and strong governance of public services. Similarly, it has a highly developed and regulated system of public spending involving mechanisms that are subject to varying degrees of parliamentary scrutiny to ensure democratic accountability and transparency in decision-making.

Public spending in the United Kingdom is generally no different from other developed Organisation for Economic Co-operation and Development (OECD) countries, and it mainly relates to supporting public services such as health and care, social security, education, defence, transport, and housing. Total Capital Departmental Expenditure Limits (CDEL) reported in the most recent (2025–2026) UK Government Spending Review (SR) are set at £111.3 billion, which accounts for 1.8% average annual real-term growth to 2029/2030 (the full SR period). Defence and transport projects account for the greatest increases in CDEL, with forecasted increases of 7.3% and 3.9%, respectively.

Governance Frameworks for Major Public Projects: International Practices and Experiences, First Edition.
Edited by Vedran Zerjav, Morten Welde and Gro Holst Volden.

The United Kingdom has a rich history of public investment projects, mainly as capital projects delivering assets within the public service delivery. The oversight of major government projects and programmes lies with the central government, whereas departments are usually responsible for delivery. Transport infrastructure, such as highways and high-speed rail, is delivered through separate 'arms-length bodies'. At the regional and local levels, there are numerous delivery models available to local government bodies such as 'county councils' and 'combined authorities'. In summary, the landscape of project delivery in the public sector is complicated.

The UK context is world-renowned for delivering a vast range of service transformation programmes, which include megaprojects (often defined in terms of £1 billion+), which comprise very large investments from the public sector to deliver major infrastructure. Similar to many other nations, infrastructure investment is a key enabler of the United Kingdom's ambitions for economic growth and prosperity. The system currently in place is a result of development and adaptation to different trends and shifts in politics, society, and finance. For example, the current geopolitical shifts arising from contested spaces and global trade 'wars' have led to a radical 'rethink' of the UK's national security and resilience needs, and the role of industrial strategy in enabling them. The challenges facing the UK infrastructure delivery sector are long-standing and well documented. Critics have emphasized how infrastructure projects are 'too costly' and lack strategic alignment with the UK's needs. Moreover, while political cycles influence decisions around major projects, economic cycles influence the availability of materials, labour, and supply chains and thus outturn costs.

Given the relatively high levels of transparency and independent media coverage, as well as the often international profile of UK planning and engineering practices, large-scale public investment projects are often seen as case studies for international practices in project planning and implementation. Recent examples include Crossrail, a fast underground West–East link in London, the 'super sewer' Thames Tideway Tunnel, and the extremely controversial High Speed Two. The United Kingdom hosts a highly developed public sector, where, similar to other developed countries, the civil service implements investment decisions independent from political parties. UK public investment projects are typically implemented through a host of actors in a mature institutional setting based on high levels of transparency, trust, and decisions made based on expertise and democratic governance.[1] Governance of major public investment projects in the UK setting revolves around the Government Major Projects Portfolio (GMPP) and the National Infrastructure and Service Transformation Authority (NISTA), which is an independent body advising the government on the performance of the projects on the GMPP.

The **GMPP** is a selection of major projects that are deemed as 'complex and/or strategically significant'. In 2024/2025, the GMPP contained 213 projects across 20 departments, with a total whole life cost of £996 billion, with forecast whole life monetized benefits of £742 billion (NISTA 2025). Each government major project on the GMPP is categorized in one of the four groups:

1 For a more comprehensive view of the assurance and governance landscape in the United Kingdom, readers are advised to refer to Vo et al. (2021).

- **Infrastructure and construction projects:** projects to improve and maintain the UK's critical infrastructure, e.g. energy, environment, transport, telecommunications, and construct new public buildings. Usually, high investment projects are regarded as essential to the nation's economic growth and prosperity and given cross-government priority.
- **Government transformation and service delivery:** projects to change ways of working and improve the relationship between government and UK citizens, including channelling new technologies to improve and increase efficiency of government/public services.
- **Military capability projects:** delivering projects to enable the effective operation of the Armed Forces to support the maintenance of the UK's national security.
- **ICT:** projects to enable movement from old legacy systems to new digital solutions, to equip departments for the future, deliver efficiencies, and harness the benefits from advancements in innovation and technology (NISTA 2025).

The NISTA sits within government and is overseen jointly by His Majesty's Treasury (HMT) and the Cabinet Office (CO). It is designed to advise ministers on the go/no-go decisions for major projects and infrastructure. NISTA is a new body, established in April 2025[2] as a merger of the former (1) National Infrastructure Commission (NIC), an independent body that assessed the government's progress on infrastructure strategy and delivery plans, primarily through the 'National Infrastructure Assessment' and (2) the Infrastructure and Projects Authority (IPA), which reported to HMT and the CO and supported the delivery of major projects in the United Kingdom through leadership of the government project delivery profession and oversight of the assurance of the major projects and programmes.[3] NISTA is tasked with overseeing GMPP. NISTA's mandate is to ensure independent quality assurance of the largest projects and to report on, support, and develop expertise in the management of large projects within the public administration. In addition, it also forms a key part of a wider governance and assurance structure and framework within government. This framework comprises varied additional approval steps, taking the form of design, commercial, and spending controls, that must be followed by UK government departments depending on the type of project being delivered.[4]

---

2 See the 10-year National Infrastructure Strategy: https://www.gov.uk/government/publications/uk-infrastructure-a-10-year-strategy.

3 The IPA was formed in January 2016 by merger of the Cabinet Office's MPA and HM Treasury's Infrastructure UK (iUK). Prior to the IPA, the MPA was primarily responsible for the governance and assurance processes underpinning major project delivery whereas iUK primarily provided expertise in financing, procurement, and various forms of public–private partnership. (This unit had been strengthened in previous years to improve procurement skills across government.) With the merger, expertise in delivery, assurance, and financing of major public projects was brought together in a single body.

4 It is important to note that, as far as governance and assurance are concerned, government departments retain the authority to conduct assurance and governance for projects that do not sit on the GMPP. This is supported by external (to the department) advice through organizations such as the GIAA. GIAA is set up to provide independent, objective assessment of governance, risk, management and control arrangements and provide advice on how to improve them. In the context of projects and programmes, this includes 'risk-based audit' on the effectiveness of risk management processes, in relation to the risk appetite and 'programme assurance' on the effectiveness of programme and project management for the delivery of benefits.

Oversight of the GMPP is based on the appraisal methodology for policies, programmes, and projects in the UK public sector set out in the Government's so-called 'Green Book'. The Green Book is underpinned by a 'five-case model' of appraisal, including strategic, economic, commercial, financial, and management dimensions that HMT deem key to public investment decision-making. The Green Book guidance applies to most aspects of public spending, taxation, and regulatory change, as well as to alternative uses of existing public assets and resources. It provides the foundation for the analysis of proposed projects and their subsequent approvals.

4.2 The UK Project Governance Model

The UK major project governance model is supported by the 'three lines of defence' approach to risk assurance (HM Treasury 2012, pp. 5–7). The 'three lines of defence' are:

- **First line:** This comes directly from those responsible for delivering specific objectives or operations.
- **Second line:** This work is associated with oversight of management activity. It is separate from those responsible for delivery, but not independent of the organizations management chain.
- **Third line:** This relates to independent and more objective assurance.

These three lines continue to be applied and form part of assurance governance structures providing 'through a systematic set of actions, confidence to senior leaders and stakeholders that work is controlled and supports safe and successful delivery of policy, strategy and objectives' (Infrastructure and Projects Authority, Cabinet Office 2021a, p. 11). This approach to assurance is seen throughout all projects, with the third line of defence – independent assurance – being a mandated requirement for all programmes and projects on the GMPP that are undertaken following a set of principles and procedures. Specifically, GMPP projects undergo a governance scheme of mandatory quality assurance at important transitions between project phases, making it effectively a stage-gate process model. This applies to the most complex, high-risk projects and programmes and is, in addition to the requirements and processes, applied by individual ministries, agencies and their Arm's Length Bodies (ALB), including spend controls.

4.2.1 A Brief Historical Overview

In 2011, the Major Projects Authority (MPA) was established in the UK Government's CO; its purpose was to provide 'oversight' of the central government's most complex and high-risk projects (the GMPP). The creation of the MPA was primarily in response to a National Audit Office (NAO) report in 2010, which called for a 'central, mandatory system of assurance to be established for government'. The report recognized the value of the pre-existing Office for Government Commerce (OGC) Gateway™ review methodology and the creation of a Major Project Review Group (MPRG) to provide scrutiny, but recommended a new system that is evidence-based when judging whether the key elements of a successful project are in place and effective.

GMPP governance is underpinned by an Integrated Assurance Strategy (IAS) and the associated Integrated Assurance and Approval Plan (IAAP). It is recommended that central government bodies develop an IAS to ensure consistency in the application of assurance standards across all GMPP projects within their department. Responsibility for the IAS lies with the Senior Responsible Owner (SRO) for each major project, and it should provide a strategic purview of the approach to assurance. The IAAP, on the other hand, is more tactical in nature and designed to align assurance activities in a way that is well-organized, timed, and coordinated with the resources available. It is usually the case that an IAAP is a prerequisite to submission of the Strategic Outline Business Case (SOBC).

In general, the purpose of assurance is to:

- Determine Delivery Confidence Assessment (DCA)
- Provide findings and recommendations to the SRO
- Provide an assurance report to inform decisions made by Government departments and HMT

The assurance reviews (sometimes known as stage-gate reviews) are conducted by a small team of trained individuals who have project delivery experience and have knowledge or expertise of at least one 'core business area', including change management, construction, estates, digital and technology, and operational delivery/management (Infrastructure and Projects Authority, Cabinet Office 2016, p. 4). The outcome report provides the SRO of the project with a series of recommendations based on the observations of the Review Team during the review. The purpose of any recommendations is to improve the likelihood of the project's successful delivery. Stage-gate assessments use a red-amber-green (RAG) rating system, where green indicates high confidence in the successful delivery of the programme/project.

4.2.2 The Stage Gate Review Model, Phases, and Decision Points

For projects and programmes on the GMPP, stage-gate review outcomes are shared with Accounting Officers (AOs), HMT, and CO leadership to support successful delivery. Stage-gates are aligned to the 'Five Case Business Case Model' and seek to examine programmes and projects at key decision points in their life cycle to provide assurance that they can progress successfully to the next stage. Refer to the Government's Project Delivery Functional Standards for more information on the project delivery life cycle and key decision points. Assurance activities are organized in accordance with the reference project life cycle described in HM Government's 'Functional Standard for Project Delivery' (Figure 4.1) and the current guidance on Assurance Review Options (Table 4.1) published by NISTA's predecessor body, the IPA.

Gate Review 0: This is applied at the front-end of the project or programme and subsequently repeated at appropriate decision points during the life cycle up until close-out.

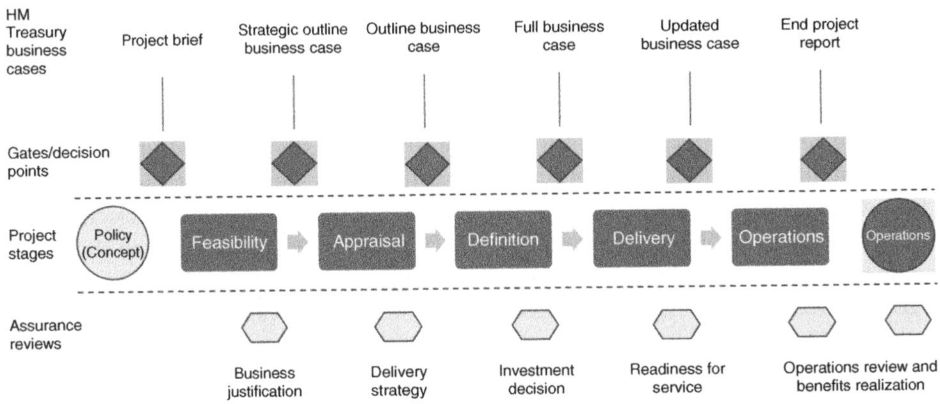

Figure 4.1 The reference project life cycle (HM Government Functional Standard for Project Delivery).

Table 4.1 Planned assurance options for inclusion in Integrated Assurance and Approvals Plans. Adapted from [6].

Project Validation Review (PVR)	Gate 0	Gates 1–5	Project Assessment Review (PAR)
Informs departmental discussions about viability of the potential project. SRO actions the findings and the recommendations to design, scope, and develop a project. HM Treasury Spending Team will expect to see the report at the next approval point.	SRO actions the findings and the recommendations. HM Treasury Spending Team will expect to see the report at the next approval point. A poor Delivery Confidence Assessment rating may, as a consequence, trigger further investigation, support, and/or intervention from the IPA.	SRO actions the findings and the recommendations. HM Treasury Spending Team will expect to see the report at the next approval point. A poor Delivery Confidence Assessment rating may, as a consequence, trigger further investigation, support, and/or intervention from the IPA.	SRO actions the findings and the recommendations. HM Treasury Spending Team will expect to see the report at the next approval point. If the project is subject to MPRG Panel scrutiny, the report will inform MPRG Panel meeting discussions and lines of questioning. A poor Delivery Confidence rating may, as a consequence, trigger further investigation, support, and/or intervention from NISTA.

This is to ensure that the business case remains aligned with the outcomes and intended benefits. It may also be required in situations where the policy landscape has changed significantly.

Gate Review 1: Business justification: This investigates the SOBC and 'a proposed way forward to confirm that the project scope is clear and achievable, and the project is likely to deliver what is required'.

Gate Review 2: Delivery strategy: It tests assumptions in the Outline Business Case (OBC) including delivery approach and procurement options.

Gate Review 3: Investment decision: The Full Business Case (FBC) and the governance arrangements for the investment decision are assessed alongside the implementation plans.

Gate Review 4: Readiness for service: This is an assessment of readiness to transition from the temporary organization (the project) to the permanent organization (operations).

Gate Review 5: Operations review and benefits realization: This confirms that the benefits set out in the Business Case are being achieved and that the operational service (or facility) is running smoothly, and the agreed strategic outcomes are being met.

At the time of writing, documentation relating to the Project Validation Review (PVR) and Project Assessment Review (PAR), as referred to in Table 4.1, has been withdrawn; it is assumed that these are superseded by the revised stage-gate 0 and stage-gate 5 guidance described earlier.

4.2.3 Business Cases and HM Treasury's Green Book[5]

The primary guidance used by government departments to inform investment decisions is the Green Book; it describes how to appraise policies, programmes, and projects and includes guidance on the design and implementation of monitoring and evaluation before, during, and after the policy intervention. This is particularly relevant to benefits realization since it provides the evidence base necessary to understand if the policy is successful in the context of its' business case.

The 'Business Case' is a summary of the findings from the analyses and appraisals made of an investment at a specific point in time. Its contents largely correspond to those of an economic analysis but may include other considerations. HMT requires the application of the so-called Five Case method, which comprises the following five key components: the strategic case, the economic case, the commercial case, the financial case, and the management case.

The Business Case is developed over time, and in an iterative way, this is to accommodate the HMT-approval process, which involves three stages, requiring the following variants of the Business Case, respectively:

- **Strategic Outline Business Case (SOBC):** The focus is primarily on strategic (and economic) issues, as well as the selection of the level of ambition. Success criteria need to be defined as a basis for filtering out alternatives. The purpose of the SOBC is to reduce the opportunity space from a long list to a short list (i.e. to bring it down to only three to four alternatives).
- **Outline Business Case (OBC):** This involves a more detailed analysis of the three to four alternatives, with a view to recommending a concept. Economic considerations are key, but one will also address the structuring of the project and

5 Based on Guide to developing the project business cases, HM Treasury (2018), and The Green Book, Central Government Guidance on Appraisal and Evaluation (2022).

the implementation approach, including the robustness of the plans, the funding opportunities, and the contract strategy.

- **Full Business Case (FBC):** This is the most comprehensive version of the business case and pertains to the chosen concept only. One needs to ensure that the project remains relevant and profitable, as well as present specific implementation plans. It needs to be documented that cost and risk are under control, that adequate know-how is available, and so on. The FBC underpins the actual investment decision.

A core feature of Green Book is the focus on the appraisal of alternative policy options – this is described in the guidance as 'an inseparable part of detailed policy development and design'. The strategic dimension of the five-case model should, therefore, describe the prevailing situation without any further action, known as 'business as usual', to determine the standard counterfactual against which alternative options are usually compared under the Green Book principles.

The Green Book is not a mechanical or deterministic decision-making device. It describes approved models of thinking and methods/methodologies to support the provision of advice to clarify the social – or public – welfare costs, benefits, and trade-offs of alternative implementation options for the delivery of policy objectives. What this means in practice is that the Green Book is designed to provide objective advice by civil servants to decision-makers, which in central government means advice to ministers – following the doctrine of 'Officials advise, Ministers decide'.

4.2.4 Projects That Must Adhere to the Model

There is typically no given budget envelope that determines inclusion of a project in the GMPP. However, the IPA states that 'the criteria for GMPP projects are typically those where approval is required from HMT, either because the budget exceeds a department's delegated authority level or because the project is novel, complex, contentious, or requires primary legislation. While the GMPP spans many of the government's most high-profile projects, it represents only a portion of the projects delivered across government, with most departments having further projects and programmes within their portfolios that do not meet the above GMPP criteria'.

As mentioned earlier, in the GMPP projects of the following types are represented: infrastructure and construction, government transformation and service delivery, military capability, and ICT. In contrast to the Norwegian governance scheme, there is a relatively higher degree of discretion to categorize a project within the GMPP in the United Kingdom.

4.2.5 Selected Topics

Alongside the governance scheme, there are several other key areas and government documents published relating to the delivery of major government projects, broadly supporting the governance scheme. These relate to the general guidance supporting delivery of major projects in government (The Project Routemap), focusing on the delivery of public construction projects and programmes (The Construction Playbook), and defining the job roles, capabilities, and learning available for those in the government project delivery profession (IPA Project Delivery Capability Framework).

4.2.5.1 The IPA 'Project Routemap'

The 'Project Routemap' promotes consideration of two perspectives that it argues are crucial to effective governance frameworks:

- Governance – 'the rules established for prioritising, authorising, directing, empowering and overseeing management, and assuring and reviewing performance'.
- Governing – 'the ongoing culture, behaviours, values and relationships, which enable effective governance'.

The guidance emphasizes that 'weak governance' can compromise the delivery of intended project benefits, including economic, environmental, and social value. Crucially, the 'Project Routemap' guidance explicitly states that arrangements (during the project/programme life cycle) 'must align with the wider and long-term corporate governance of the sponsor organisation' (Infrastructure and Projects Authority, 2014). In practice, this suggests that governance must recognize the complex risks that are associated with government projects that are promoted and approved on the basis of delivering benefits in ways in which the government delivers services, as well as to the public who are the ultimate recipients of those intended benefits.

4.2.5.2 IPA Project Delivery Capability Framework

The IPA, alongside coordinating the assurance review of major projects and establishing project delivery functional standards, has provided a Project Delivery Capability Framework (Infrastructure and Projects Authority, Cabinet Office 2021c). The Framework defines the job roles, capabilities, and learning available for those in the government project delivery profession. It also outlines the skills, experience, and process to follow to obtain accreditation as a Government Project Delivery Professional (Infrastructure and Projects Authority, 2021c). The accreditation process is supported by the Government Online Skills Tool, which enables its users (government project professionals) to assess their skills, identify gaps, provide learning solutions to address them, and build personal development plans. It also enables teams and departments to identify gap areas and develop specific capability initiatives.

4.2.5.3 The Construction Playbook

The *Construction Playbook*, (Cabinet Office, 2022), sets out 14 key policies and provides guidance on the delivery of public construction projects and programmes. The guidance includes the role of the SRO, who owns the business case and is accountable (to Parliament) for the delivery of the project or programme and its intended benefits and outcomes. The guidance notes that SROs 'should fully understand the governance and approvals process and commit sufficient time to lead the project or programme through approvals and delivery'. The 'playbook' describes the importance of ensuring the SROs ability to 'lead teams of people with the necessary mix of functional expertise and experience to match the capabilities of the market'. In practice, this can be challenging given the need to appoint a person who is sufficiently knowledgeable of the business case and able to influence programme and project teams in resolving tensions between project performance metrics and project success criteria (see Bosch-Rekveldt et al. 2023).

4.2.5.4 Post-project Reviews

There is a requirement for a formal external assurance review to be completed once the project has become operational, i.e. it has delivered what was agreed to contribute to its outcomes, which could be, for example, new information technology, estate, or a training programme. This Gate 5 review will look at both operations and benefits realization, the first of which will coincide with any primary decision points following the project's internal post-implementation review. The suggested best practice is for Gate 5 reviews to be conducted several times throughout the project's likely long-term operational service, with the first focusing on project closure, lessons learned, transition to 'business as usual', business case and benefits monitoring, and delivery. Mid-stage reviews will also focus on benefits realization, and the final review will centre on service arrangements and lessons learned (Infrastructure and Projects Authority, Cabinet Office 2021b).

4.3 Experience with the Governance Scheme

Although it is difficult to summarize the experience of running the UK major projects governance scheme due to its complexity and entanglement with various other bodies and processes,[6] the recently published NISTA Annual Report on its GMPP (2024–2025) presents a summary of the shift between DCA ratings in projects, which potentially provides an indication of the effectiveness of the reviews themselves. DCAs indicate the level of confidence in the project or programme's ability to deliver its aims and objectives within the timescales and cost envelope agreed and to the quality requirements, including financial and non-financial benefits, as agreed in the business case. For example, of the 178 projects that remained on the portfolio and were subject to reviews between 2022/2023 and 2023/2024, only 22 improved their DCA, with 142 remaining the same and 14 becoming worse. It was acknowledged there had been an increase in the number of 'red-rated' projects attributed to two primary factors: new projects joining the GMPP and 'unprecedented issues over recent years such as high inflation, challenging net zero targets, global demand and skills shortages' (NISTA 2025).

In an independent empirical study of the UK GMPP assurance data, Vo et al. (2021) discuss the evidence for the effectiveness and limitations of project assurance reviews. They find that the UK's project governance scheme, as implemented through the GMPP and its assurance processes, is conceptually robust and structurally sound, offering a valuable framework for oversight, learning, and accountability. However, its practical effectiveness is undermined by inconsistent implementation of recommendations, weak enforcement, and follow-up mechanisms, limited focus on strategic outcomes and benefits and organizational and cultural barriers to learning. As a result, the authors suggest the following recommendations to improve the practices of assurance.

- Mandate Departmental Portfolio Officers (DPOs) to track implementation.
- Improve reviewer training and ensure continuity in review teams.

---

6 In particular, relating to the effectiveness of the advice and governance measures introduced by the different bodies that were put in place for this purpose over the years (Major Projects Association, Infrastructure and Projects Authority, Infrastructure UK).

- Balance focus between delivery and benefits in recommendations.
- Limit the number of recommendations per review to critical issues.
- Encourage SRO engagement in defining review scope and acting on findings.

At a more anecdotal level, it is important to mention that the IPA 2022–2023 annual report assigned an 'unachievable' rating for the delivery confidence rating for phases 1 and 2A of the High Speed Two, playing a significant role in the subsequent cancellation of the project, announced by PM Rishi Sunak in October 2023. HS2, a programme to deliver high-speed rail between London and Birmingham, has been mired in controversy and criticism of its governance arrangements. The cancellation, following the publication of the IPA annual report and its delivery confidence rating, can be seen as an example of the governance scheme making an impact on the ongoing projects.

To conclude, while some solid indications are pointing towards the effectiveness of the scheme, it would require a more comprehensive follow-up study to examine the interplay of the project delivery performance with the institutional arrangement put in place to enact the project governance scheme.

References

Bosch-Rekveldt, M., Bourne, M., Forster, R. et al. (2023). Performance measurement in project management. *International Journal of Project Management* 41 (7): 102518. https://doi.org/10.1016/j.ijproman.2023.102518.

Cabinet Office. (2022). The Construction Playbook: Government guidance on sourcing and contracting public works projects and programmes (Version 1.1) [PDF]. GOV.UK. https://assets.publishing.service.gov.uk/media/6312222de90e075880923330/14.116_CO_Construction_Playbook_Web.pdf.

HM Treasury (2012). *Assurance Frameworks*. London, UK: Crown Copyright.

HM Treasury (2018). Guide to developing the project business case: Better business cases—for better outcomes. https://assets.publishing.service.gov.uk/media/66449468ae748c43d3793bb8/Project_Business_Case_2018.pdf (accessed 17 December 2025).

HM Treasury (2022). The Green Book: Central government guidance on appraisal and evaluation (2022 update). https://assets.publishing.service.gov.uk/media/6645c709bd01f5ed32793cbc/Green_Book_2022__updated_links_.pdf (accessed 17 December 2025).

Infrastructure and Projects Authority. (2014). Project Routemap: Setting up projects for success. https://www.gov.uk/government/publications/improving-infrastructure-delivery-project-initiation-routemap (accessed 17 Dec 2025).

Infrastructure and Projects Authority (2017). Guidance for Departments _A guide to choosing which IPA Assurance Reviews to include in the integrated Assurance and Approvals Plan for your Major Project (no URL available).

Infrastructure and Projects Authority, Cabinet Office (2016). *Becoming an Assurance Reviewer for the Infrastructure and Projects Authority*.

Infrastructure and Projects Authority, Cabinet Office (2021a). *Government Functional Standard GovS 002: Project Delivery*. London: Crown Copyright.

Infrastructure and Projects Authority, Cabinet Office (2021b). *Gate Review Process Gate 5 Review: Operations Review and Benefits Realisation*. London: Crown Copyright.

Infrastructure and Projects Authority, Cabinet Office (2021c) *Project Delivery Capability Framework*. https://assets.publishing.service.gov.uk/media/65561f36046ed4000d8b9a33/PDCF-V3.pdf(accessed 9 March 2025).

Infrastructure and Projects Authority, Cabinet Office (2025). *Annual Report on Major Projects 2023–24*. London: Crown Copyright.

National Infrastructure and Service Transformation Authority. (2025). NISTA Annual Report 2024–2025. https://www.gov.uk/government/publications/nista-annual-report-2024-2025/nista-annual-report-2024-25 (accessed 17 Dec 2025).

Vo, H., Kirkham, R.J., and Williams, T.M. (2021). An empirical study of assurance in the UK government major projects portfolio: from data to recommendations, to action or inaction. *International Journal of Managing Projects in Business* 14 (4): 865–897.

5

Ireland

Garret Doocey[1] *and Ed Hearne*[2]

[1] *Department of Transport, Dublin, Ireland*
[2] *Fiscal Affairs Department, International Monetary Fund, Washington, DC, USA*

Over the past decade, Ireland has significantly increased its public investment to address long-standing infrastructure deficits and support sustainable national development. This shift has been accompanied by a series of reforms aimed at strengthening the governance of major public projects. Central to these reforms is the publication of the *Infrastructure Guidelines* in 2023, which consolidate and build upon earlier frameworks to ensure more rigorous project appraisal, oversight, and delivery (Department of Public Expenditure, Infrastructure, Public Service Reform and Digitalisation, 2023a). This chapter outlines the evolution, structure, and operational mechanisms of Ireland's project governance model, highlighting how it seeks to enhance accountability, efficiency, and value for money in public investment.

5.1 Context and Policy Basis

Ireland is a small, open economy with a population of 5.4 million in 2024 (Central Statistics Office 2024). Ireland has been a member of the European Union (EU) since 1973 and experienced robust economic growth from the 1990s, driven by foreign investment. In the wake of the 2008 Financial Crisis and Great Recession, Ireland underwent a significant economic contraction but has been among the fastest-growing advanced economies in the world in the past decade. The State is unitary and has a level of decentralization to regional and local government among the lowest in the EU. Perceptions of corruption are comparatively low (Transparency International 2024), and the share of the population with moderate or high trust in public institutions such as the national government, the judiciary, the civil service, and parliament ranks above the Organisation for Economic Co-operation and Development (OECD) average (OECD 2024).

The views expressed in this chapter are those of the authors and do not necessarily represent the views of the DoT or the IMF.

Governance Frameworks for Major Public Projects: International Practices and Experiences, First Edition.
Edited by Vedran Zerjav, Morten Welde and Gro Holst Volden.

Public investment has been volatile over the past 30 years. Government capital spending has been pro-cyclical, increasing in times of economic growth and decreasing during contractions (Conroy et al. 2021). The public capital stock – a measure of the total value of physical assets owned and maintained by the government, including economic and social infrastructure – has fluctuated as a result and is lower than in other high-income countries both as a share of national income and in per capita terms (Conroy and Timoney 2024). In addition, analysis has shown that the quality of Ireland's public infrastructure is lower than that of other advanced economies with comparable levels of capital stock per capita, signalling an 'efficiency gap' in infrastructure spending (Renteria et al. 2017).

Over the decade since 2015, policy has sought to ramp-up public investment and bridge infrastructure deficits. To illustrate, the central government capital budget for 2025 was €14.9 billion – over €10 billion higher than the 2016 outturn (Government of Ireland 2018a, 2024). Successive updates to Ireland's National Development Plan (NDP) have detailed the projects and programmes to be pursued across transport, energy, water and sanitation, education, healthcare, housing, and cultural infrastructure (Government of Ireland 2018a, 2021). Policy has also sought to ensure that infrastructure investment is aligned with a strategic spatial strategy for national development (Government of Ireland 2018b) and in line with climate action targets (Government of Ireland 2021, 2023).

In tandem, a series of steps has been taken to enhance investment preparation. These have included revisions to arrangements for project appraisal and selection, new structures for oversight of the implementation of national infrastructure policy, and an overhaul of arrangements for governance of major projects. The *Infrastructure Guidelines*, published in 2023, reiterate commitments to long-standing practices for investment appraisal, further embed arrangements for major project governance and update roles and responsibilities of key players.

5.2 Project Governance in Ireland – The Infrastructure Guidelines

5.2.1 History and Development of the Model

The *Infrastructure Guidelines* were published in December 2023, although in practice most provisions therein have been in place for some years. The Guidelines apply to all public bodies and all bodies in receipt of public funding for capital projects and programmes.[1]

The roots of the project governance model in Ireland can be traced to the requirements for cost–benefit analysis of grant-aid proposals for industrial policy development in the 1970s. In the early 1980s, requirements for appraisal were introduced for all government capital projects. The requirements were further codified with the publication of the Capital Appraisal Guidelines in 1994. The Capital Appraisal Guidelines established a loose stage-gate process with four discrete phases of the project life cycle: appraisal, planning and design, implementation, and ex post review and were updated in 2005

1 Throughout this chapter, the terms *project* and *programme* are used interchangeably.

(Department of Finance 2005). The emphasis was on the appraisal phase and technical issues in economic appraisal of projects rather than issues relating to design, procurement, and implementation.

The project governance system evolved over the 2000s. Within the overall framework provided by the Capital Appraisal Guidelines, line ministries established sector-specific frameworks.[2] The publication of the *Public Spending Code* (Department of Public Expenditure and Reform 2013) sought to consolidate the various requirements. The Code was further updated in 2019, focusing more on risk management. New assurance processes for major projects were added in 2021, partially inspired by the Norway model. Finally, 2023 saw the publication of the *Infrastructure Guidelines*, which maintain the thrust of the Public Spending Code and associated assurance processes while seeking to streamline the approval arrangements for major projects.

5.2.2 How Projects Are Initiated

Project proposals are typically initiated by the relevant government ministry or agency with sectoral responsibility (the 'Sponsoring Agency'). In most sectors, the origins of major projects can be traced to an overarching sectoral strategy. This is usually consistent with wider regional policy and in line with the NDP. Proposals can also be initiated by local authorities and other public bodies. In common with many countries, broad concepts can emerge years, or even decades, before they are formalized as project proposals.

5.2.3 The Stage-Gate Model with Phases and Decision Points

The model consists of a five-stage project life cycle model as depicted in Figure 5.1.

1) Strategic Assessment and Preliminary Business Case
2) Project Design, Planning and Procurement Strategy
3) Post-Tender – Final Business Case
4) Implementation
5) Post Completion Review and Benefits Realization.

The *Guidelines* adopt a proportionate approach in relation to their application with three broad categories of projects/programmes acknowledged:

- Projects under €20 million
- Projects between €20 million and €200 million
- Projects over €200 million, the threshold for designation as a major project

For major projects, the *Guidelines* require Government approval at Approval Gates 1 (AG1) and 3 (AG3), with ministerial approval a requirement at Approval Gate 2 (AG2).

2 For example, in industrial development (2005), transport investment (2004, 2023), water services (2009), and later higher education facilities (2019) and cultural infrastructure (2021).

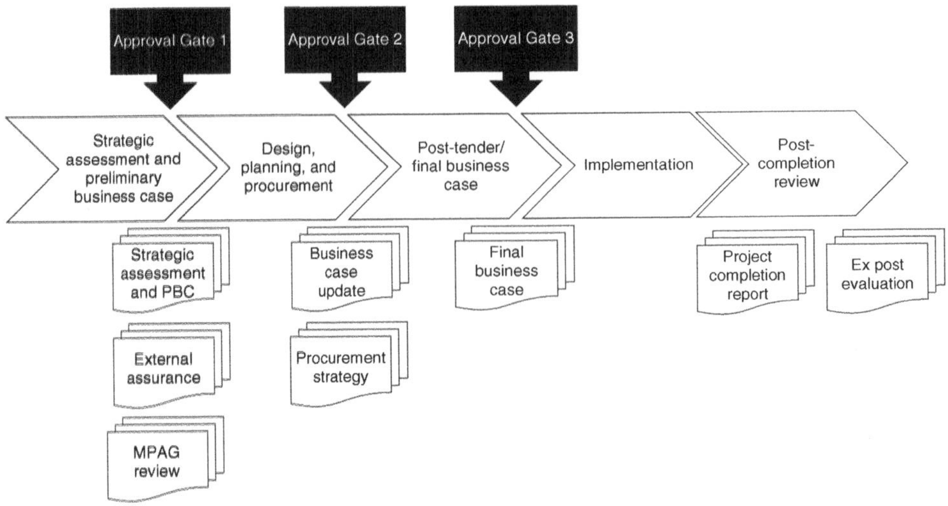

Figure 5.1 Infrastructure guidelines project life cycle.

5.2.3.1 AG1 – Strategic Assessment and Preliminary Business Case

At AG1, project proposals should demonstrate the rationale for the preferred option and approach to delivery and outline the preliminary design upon which the costings and economic and financial appraisals are based. In the transport sector, AG1 occurs later in the project life cycle, typically before entry into the statutory planning and permitting process, when more detailed analysis is available for consideration. The Approving Authority must assess the project's Preliminary Business Case in order to:

- Test assumptions in the business case.
- Validate cost and schedule estimates.
- Analyse the treatment of risk.
- Consider other issues such as stakeholder management and benefits realization.

For major projects, the Approving Authority then forwards the project documentation to the Department of Public Expenditure, Infrastructure, Public Service Reform and Digitalisation (DPER), which then convenes a meeting of its Major Projects Advisory Group (MPAG) to review the proposal. Major projects are then submitted by the line ministry to the government for its approval to proceed to AG2.

5.2.3.2 AG2 – Project Design, Planning, and Procurement Strategy

AG2 is pre-tender approval and allows the preferred option, as identified at AG1, to be further reviewed as the project's underpinning assumptions are refined, its risks better understood, and the design stage is sufficiently developed to prepare the Final Business Case. Currently at AG2, major projects require approval from the relevant Minister before commencing procurement.

5.2.3.3 AG3 – Post-Tender – Final Business Case

AG3 is final approval. The purpose of the Final Business Case is to reassess the assumptions underpinning the Preliminary Business Case and reconsider the findings emerging. For major projects, government consent must be secured prior to the award of the main construction contracts.

5.2.3.4 Implementation

Implementation of a project is the responsibility of the Sponsoring Agency, while the Approving Authority (and Accounting Officer) must be satisfied that the Sponsoring Agency delivers the project in line with the approvals granted.

5.2.3.5 Post-completion Review

Project Completion and Ex Post Evaluation reporting are the final stages in the project life cycle. The Guidelines require that the first of these must be undertaken as the project is being completed to allow the reviewer access to feedback from the contractor(s) and project team. The report should be conducted by the Sponsoring Agency and submitted to the Approving Authority for review and, for major projects, is required to be forwarded to DPER. The Guidelines state that the Ex Post Evaluation Report should be conducted by the Sponsoring Agency, typically between three and five years after project completion and should examine whether the expected benefits and outcomes of the project materialized as forecasted and what conclusions can be drawn to inform future projects. As with the Project Completion Report, the report is required to be submitted to the Approving Authority and, for major projects, onwards to DPER.

5.2.4 Projects That Must Adhere to the Model

The rigour and coverage of the quality assurance process are commensurate with the scale and complexity of the project or programme under consideration. The largest and most costly investments are subject to closer scrutiny and assessment of costs, benefits, and risks at various points in the project life cycle. Major projects, classed as those projects with estimated costs in excess of €200 million, are subject to review by MPAG, while smaller and more straightforward investment proposals receive a proportionate degree of examination.

The model applies to all publicly funded investment projects and programmes, including central government, local authorities, and other public bodies. Major investments by commercial state-owned enterprises, such as electricity generation and transmission or natural gas network assets, are not covered by the model.

5.2.5 Institutions and Roles

The Infrastructure Guidelines set out the roles and responsibilities in relation to the various actors in the development of major projects. There are three key administrative roles specified within the Guidelines:

- The Accounting Officer
- The Approving Authority
- The Sponsoring Agency

In addition, both the DPER and the Government retain important responsibilities for major projects.

5.2.5.1 The Accounting Officer

The Accounting Officer is typically the Secretary General of the relevant ministry, and their statutory duties are set out in law and administrative regulations. A key function of the Accounting Officer is their personal responsibility '*for the safeguarding of public funds and property under his or her control*' and in relation to public investment projects, the

Infrastructure Guidelines state that the Accounting Officer of the ministry funding a project has ultimate responsibility for the project.

The references to the Accounting Officer are the only such references to a specific individual postholder within the *Infrastructure Guidelines*, and there is no explicit equivalent to the role of Senior Responsible Owner as in other jurisdictions such as Australia, New Zealand, or the United Kingdom.

5.2.5.2 The Approving Authority[3]

The Approving Authority is typically the ministry funding the project; however, the *Infrastructure Guidelines* note that in some sectors, such as transport, there can be delegated sanction arrangements, which result in agencies assuming a role as 'day-to-day approving authorities' on behalf of the parent ministry.

The key responsibilities of Approving Authorities are to:

- Assess all documentation submitted by Sponsoring Agencies to ensure compliance with the *Infrastructure Guidelines*.
- Prepare sectoral-specific guidance for evaluating, planning, and managing capital expenditure.
- Seek government approval at AG1 and AG3 and ministerial approval at AG2.
- Monitor the project as it is implemented and review whether the project should progress in instances where major developments occur that threaten the viability of the project.
- Review the Project Completion Report and Ex Post Evaluation Report and ensure any relevant findings are incorporated within sectoral guidance.
- Notify the government of any adverse developments and advise as to potential mitigation measures.

5.2.5.3 The Sponsoring Agency

The Sponsoring Agency is the public body with primary responsibility for evaluating, planning, and managing the project. A Sponsoring Agency could be a ministry, local authority, higher education institute, cultural institution, or some other State agency.

The key responsibilities of Sponsoring Agencies are to:

- Prepare all documentation at relevant Approval Gates in compliance with the Infrastructure Guidelines and any relevant sectoral-specific guidance.
- Submit all documentation to the Approving Authority at relevant Approval Gates.
- Procure the project in line with national procurement rules.
- Ensure the project is implemented in line with the approvals granted by the Approving Authority.
- Manage the implementation of the project in line with those approvals and notify the Approving Authority immediately if developments occur that threaten the viability of the project.
- Plan and conduct a review of the project, incorporating lessons learned and submit a Project Completion Report to the Approving Authority when the project is completed.
- Plan and conduct an ex post evaluation of the project, incorporating lessons learned into processes and guidance, and submit the report to the Approving Authority.

---

3 The *Infrastructure Guidelines* refer to these responsibilities as being those of both the Accounting Officer and the Approving Authority; however, for ease of understanding and reflecting actual practice the authors have drawn a distinction here between the two roles.

5.2.5.4 Department of Public Expenditure, Infrastructure, Public Service Reform, and Digitilisation

DPER retains a number of important responsibilities regarding capital expenditure generally and major projects specifically. The National Investment Office, situated within the DPER, oversees and reports on government-wide investment under the NDP and acts as the secretariat for the MPAG.

Some of the key responsibilities are to:

- Lead the formulation of public investment policy through the NDP.
- Set the overall multi-annual capital expenditure ceiling for each ministry.
- Monitor actual aggregate capital expenditure across departments against monthly expenditure profiles submitted by ministries.
- Update and disseminate capital appraisal guidelines to ministries.
- Promote compliance with the *Infrastructure Guidelines* through the provision of appropriate guidance to ministries.
- Support the MPAG in its consideration of major projects.

The MPAG was established by DPER in 2021 to further strengthen project assurance. MPAG consists of external members with expertise in planning, managing, and delivering major public investment projects, along with representatives from DPER and the National Development Finance Agency (the government's PPP support body).

The key responsibilities of MPAG are to:

- Consider the preliminary business case and review reports for major projects at AG1.
- Advise on potential issues and risks related to major project costs, schedules, and benefits based on a review of the preliminary business case and assurance review report.
- Consider potential reforms to the Infrastructure Guidelines and governance arrangements for major projects.

5.2.5.5 Government

Major projects require explicit government consent to proceed at two stages of the project life cycle – AG1 Strategic Assessment and Preliminary Business Case and AG3 Final Business Case. There is no parliamentary role in the major project approval process, but issues in relation to the delivery of major projects are subject to scrutiny through normal parliamentary channels. It is the responsibility of the relevant Approving Authority (or line ministry where the Approving Authority is not a government ministry) to seek government consent through a Memorandum for Decision.

Figure 5.2 represents the parties and roles in the process.

5.2.6 Option Appraisal

The model requires assessing multiple options for achieving the desired policy objectives, with the range of options to be considered progressively narrowing as the proposal matures. The Strategic Case for the proposal should set out the long-list of options that can achieve the desired outcomes and should include solutions that do not involve capital investment where relevant. Guidance documents note approaches to developing options (such as use of workshops) and approaches to assessing the long-list through a common framework (such as a balanced scorecard, multi-criteria analysis, and programme logic frameworks). The Strategic Case should help determine the short-list of options to be analysed in detail

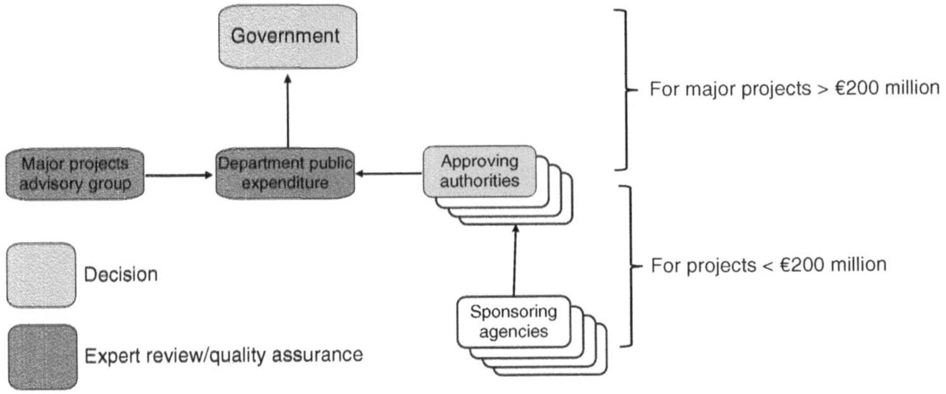

Figure 5.2 Parties and roles in the governance model.

in the Preliminary Business Case. The short-listing process should take into account factors such as affordability, scope, and deliverability. The Guidelines establish the type of appraisal required, including economic and financial appraisal and sensitivity analysis. Based on this analysis, the Preliminary Business Case should present the preferred option to be taken through to design and planning.

Sponsoring agencies are required to present an up-to-date appraisal of the preferred option in the Final Business Case, reflecting developments in cost, benefit, and delivery information resulting from the design and procurement processes. This should include updated financial and economic indicators and update the sensitivity and scenario analysis of the preferred option.

Options must be appraised with regard to a realistic counterfactual, that is, what would happen without the project. Counterfactuals can be based on a 'do nothing' or 'do minimum' scenario, but the guidance cautions against using a 'catastrophic do nothing scenario'. This occurs where doing nothing would result in major risks or adverse policy outcomes that would be unacceptable to the government. Choosing such scenarios as the investment counterfactual will artificially inflate the case for any intervention and may lead to a sub-optimum proposal being recommended.

5.2.7 Cost Estimation and Budget Management

There is no standard approach to cost estimation. The guidelines for Preliminary and Final Business Case preparation urge that particular care is taken to account for risk and arrive at robust cost estimates. While no single approach is prescribed, techniques such as benchmarking, use of the outside view and reference class forecasting (RCF) are noted. The Guidelines also note the use of International Cost Management Standards to estimate capital costs and incorporate the embodied cost of carbon associated with project proposals. Some recent megaprojects have deployed RCF as part of the appraisal, including MetroLink (Transport Infrastructure Ireland 2021), BusConnects (National Transport Authority 2022), and the Eastern and Midlands Water Supply Project (Irish Water 2024). Transport Infrastructure Ireland has also developed a framework for the use of RCF in the roads sector (see Box 5.1).

Box 5.1 Cost forecasting in the roads sector

RCF is a method to remove optimism bias and strategic misrepresentation in cost and schedule estimates for investment projects and programmes (Flyvbjerg et al. 2016). RCF emphasizes adopting an 'outside view' by analysing a reference class of similar completed projects instead of relying solely on project-specific data (Flyvbjerg 2013). Transport Infrastructure Ireland has developed guidelines for RCF to improve cost estimates for national road projects.

The guidelines outline several key recommendations for effective implementation of RCF. Forecasters are encouraged to identify a relevant reference class that closely aligns with the project under consideration. The guidelines include probability distributions based on historical data from the reference class, helping to recognize the range of potential outcomes and associated risks. The approach sets out the required uplifts to individual schemes to attain given probabilities of accuracy.

Cost estimation is undertaken iteratively as proposals mature. The guidelines state that at preliminary design, the cost estimate should have a confidence level of 50%; at the pre-tender stage, it should have risen to 65%, and finally, at the pre-award stage (after bids are received), the cost estimate should have an 80% confidence level (Transport Infrastructure Ireland 2020).

In its first end-of-year report, MPAG noted that cost estimates presented in preliminary business cases it had reviewed during 2022 did not always adequately account for risks to the scope, delivery programme, and inflation. A recent analysis by DPER assessed the causes for cost underestimation in major projects and examined measures that can be taken to more accurately account for risk in estimating major project costs (Department of Public Expenditure, Infrastructure, Public Service Reform and Digitalisation 2023b). The analysis noted the intention to update appraisal guidance to provide more detailed advice on cost forecasting and set out three areas for improvement:

- **Encourage enhanced data collection and maintenance:** The updated guidance should encourage data sharing among ministries, as well as guidance on maintaining data to ensure it is of sufficient quality to allow for the creation of appropriate reference classes and use for cost forecasting.
- **Highlight the impact of optimism bias on cost forecasts:** The updated guidance should use examples from previous projects to highlight the dangers associated with optimism bias and provide advice on how to mitigate against it.
- **Guidance on cost forecasting methodologies:** The updated guidance should provide advice about when best to use particular cost forecasting methodologies, the data requirements for such methodologies, guidance on dealing with optimism bias, and a practical guide to using particular methodologies.

5.2.8 Quality Assurance of Decision Documents

Ireland's governance model consists of tiered scrutiny of project documents ahead of critical decision points. At AG1, Sponsoring Agencies must present the Strategic Case and Preliminary Business Case (SCPBC), which must include financial and economic appraisal and should be conducted in line with detailed guidance. Following initial scrutiny by the Approving Authority, there was a requirement that the SCPBC for major projects be reviewed independently as part of the External Assurance Process prior to submission to MPAG; however, in December 2025, the government announced its intention to streamline the review process and replace the use of external consultants with in-house expertise (see Box 5.2). The review is required to consider the appropriateness of the assumptions underpinning the SCPBC, in particular relating to cost, schedule, risks, and benefits. It assesses compliance with the requirements of the Infrastructure Guidelines. At AG2, the Approving Authority reviews the Detailed Business Case and Procurement Strategy to ensure compliance with requirements. There is no formal requirement for review by MPAG at AG2, but Approving Authorities may avail of the option. In addition, at AG1, MPAG can request that proposals be resubmitted when they reach AG2, in cases where it identifies a need for further scrutiny. At AG3, the Final Business Case is prepared, updating key inputs to the appraisal with information resulting from the tendering process. At this point, all parties should have a fuller understanding of the project's cost, delivery schedule, risk profile, and benefits. The Final Business Case is reviewed by the Approving Authority. For major projects, classed as those projects with a total estimated cost in excess of €200 million, Government consent must be sought before contract award.

Box 5.2 Planned changes to the Model

In December 2025, the Government published the report and action plan of Accelerating Infrastructure Taskforce. The report sets out an roadmap designed to address bottlenecks and streamline the framework for planning, funding, and delivery key infrastructure projects and programmes. Regarding the governance model for public investment, the action plan sets out a number of key reform steps:

- Increase thresholds from €200 million to €500 million for sectors with a well-established and proven track record of delivery
- Replace External Assurance Process (EAP) with a streamlined and time bound internal review by DPER.
- Introduce administrative guidance on timelines for Preliminary Business Case (PBC) preparation (3–6 months) and mandate a maximum length.
- Allow Accounting Officers to approve projects at Approval Gate 2.
- Review and streamline the multistage approval process for transport projects.
- Assign target timeframes to all approval gates and consents for critical infrastructure projects.

Source: Government of Ireland (2025).

5.3 Experience with the Model

5.3.1 Early Lessons

While Ireland has had arrangements for project appraisal and selection in place for several decades, the strengthened oversight and assurance regime for major projects was only initiated in late 2021, and it is, therefore, too early to judge the impact of the process on project outcomes. However, a number of lessons from the early experience with the model were documented by MPAG (Department of Public Expenditure, Infrastructure, Public Service Reform and Digitalisation 2023c) and have informed amendments to the process to improve its effectiveness and efficiency. These improvements include:

- Streamlining engagement between the line ministry and the MPAG to reduce the volume of material submitted;
- A requirement that line ministries ensure that the PBC, as the PBC is submitted to MPAG, clearly highlights how the document has reflected any recommendations made by the external assurance review and if recommendations are reflected.

The process is reasonably transparent – MPAG typically publishes its Review Note after the Government has made a decision on the project and Review Notes are published on the Group's webpage. The PBC should be published as soon as possible after the government has taken the decision to approve the project.

5.3.2 Portfolio Management and Budgeting for Contingency

There is no standard approach to portfolio management across the public sector. In the roads sector, Transport Infrastructure Ireland sets the 'target scheme cost' at the P50 cost estimate, whereas the 'total scheme budget' is set at the P80 value. The agency has autonomy to manage this within its own investment allocation. Improved accuracy in cost forecasting means that under-runs against the P50 estimate in one scheme can fund over-runs in another, where they occur. This approach is aided by the composition of the agency's portfolio, which is made up of a relatively large number of similar-sized projects.

Typically, contingency is managed from within a ministry's capital expenditure ceiling or from within a day-to-day approving authority's capital allocations. For megaprojects, or in the case of ministries where a single project accounts for a major share of the total investment budget, this approach is not feasible. The issue of how best to manage contingency arrangements for certain projects, particularly megaprojects, and who is best placed to hold certain contingency levels is a topic that likely requires further consideration.

5.3.3 Developing Capacity

Supporting the effectiveness of the model will require enhancing the technical capacity of public sector professionals involved in the evaluation, planning, and implementation of public investment projects. While reforms such as the establishment of the Irish Government Economic and Evaluation Service in 2012 expanded the civil service's capacity for tasks such as economic and financial appraisal of projects, there is a need for continued focus on the range of other skills and professions required to successfully deliver major investment projects. In particular, there is likely to be an unevenness to the distribution of required skills across the public sector. Agencies with a longer history of large investment programmes have developed the required professional competencies, career structures, and project management practices and have had the opportunity to learn from past experiences. In other cases, agencies that did not historically manage major projects (including information technology investments) are now accountable for significant capital allocations. Ensuring appropriate capacity will be key to achieving value for money.

In addition, enhancing the data architecture and availability of project-level information will aid future review and evaluation of individual investments and of the effectiveness of the governance model. This should include information on cost, schedule, and benefit estimates at each point of the process and the outturn data from the project completion phase.

5.3.4 Embedding the Reforms

The *Guidelines* note that '*there is also a necessity to ensure that projects are delivered in a timely manner and that there are no unnecessary administrative delays hindering this*'. The Parliamentary Budget Office (PBO) has positively noted MPAG's review turnaround time (Parliamentary Budget Office 2024); however, what is less well understood is the length of time taken within agencies in the first instance and then between agencies, line ministries, and the external reviewers to prepare the necessary documentation for submission to MPAG. The most recently agreed Programme for Government committed towards expediting the '*delivery of project assessments, appraisals, business cases and other stages of the approvals process with a view to aiding quicker decision making*', and there have been some discussions at parliamentary committees which similarly indicate room for efficiency gains in terms of the management of overall process timelines.[4]

In December 2025, the Government published the report and action plan of the Accelerating Infrastructure Taskforce. The actions include planned changes to the governance model (see Box 5.2), which offer potential to streamline delivery but should be balanced against the risk of cutting corners in preparing and appraising the largest projects. Looking to the future, it is important that all stakeholders recognize the role and importance of front-end governance and the potency of early-stage project preparation in preventing project setbacks down the line should not be overlooked: the wealth of evidence presented in Chapters 15 and 16 underscores the importance of 'measuring once, cutting twice'. Fostering and consolidating a robust governance process can thereby assist Ireland in delivering major projects on budget and on time while achieving key social, economic, and sustainability objectives.

References

Central Statistics Office (2024). Population and migration estimates, April 2024.

Conroy, N. and Timoney, K. (2024). Ireland's infrastructure demands. Irish Fiscal Advisory Council Working Paper Series No. 23. Dublin.

Conroy, N., Casey, E., and Jordan-Doak, E. (2021). Ireland's next ramp-up in public investment. Analytical Note, Irish Fiscal Advisory Council.

---

4 See, for example, the Joint Committee on Transport and Communications, 14 February 2024; Joint Committee on Housing, Local Government and Heritage 18 June 2024.

Department of Finance (2005). Updated Guidelines for the Appraisal and Management of Capital Expenditure Proposals in the Public Sector. Dublin.

Department of Public Expenditure and Reform (2013). The Public Spending Code: Expenditure Planning, Appraisal & Evaluation in the Irish Public Service – Standard Rules & Procedures. Dublin

Department of Public Expenditure, Infrastructure, Public Service Reform and Digitalisation (2023a). Infrastructure Guidelines.

Department of Public Expenditure, Infrastructure, Public Service Reform and Digitalisation (2023b). An Analysis of Cost Forecasting in Major Capital Projects & Programmes.

Department of Public Expenditure, Infrastructure, Public Service Reform and Digitalisation (2023c). Major Projects Advisory Group – End of Year Report 2022

Flyvbjerg, B. (2013). Quality control and due diligence in project management: getting decisions right by taking the outside view. *International Journal of Project Management* 31 (5): 760–774.

Flyvbjerg, B., Hon, C.K., and Fok, W.H. (2016). Reference class forecasting for Hong Kong's major roadworks projects. *Proceedings of the Institution of Civil Engineers-Civil Engineering* 169 (6): 17–24.

Government of Ireland (2018a). Project Ireland 2040 – National Development Plan.

Government of Ireland (2018b). Project Ireland 2040 – National Planning Framework.

Government of Ireland (2021). Climate Action Plan.

Government of Ireland (2023). Climate Action Plan 2024.

Government of Ireland (2024). Budget 2025 – Expenditure Report.

Government of Ireland (2025). Accelerating Infrastructure Taskforce – Report and Action Plan.

Irish Water (2024). Water Supply Project Eastern and Midlands Region – Preliminary Business Case.

National Transport Authority (2022). BusCOnnects Preliminary Business Case.

OECD (2024). OECD Survey on Drivers of Trust in Public Institutions 2024 Results – Country Notes: Ireland.

Parliamentary Budget Office (2024). Capital Spending: An Overview of Ireland's Infrastructure Guidelines.

Renteria, C., Harris, J., Groom, S., and Allen, R. (2017). *Ireland: Technical Assistance Report-Public Investment Management Assessment*. Washington DC: International Monetary Fund.

Transparency International (2024). Corruption Perceptions Index 2023.

Transport Infrastructure Ireland (2020). Reference Class Forecasting – Guidelines for use in connection with National Roads Project. Dublin

Transport Infrastructure Ireland (2021). MetroLink – Preliminary Business Case.

6

The Netherlands

Niek Mouter

Delft University of Technology, Technology, Policy and Management Engineering Systems & Services, Delft, The Netherlands

The Netherlands has developed a distinctive governance framework for major infrastructure and water management projects, centred around the Multi-Year Plan for Infrastructure, Spatial Planning and Transport (MIRT). Unlike other countries with unified systems across ministries, Dutch infrastructure governance is decentralized, with each ministry operating its own procedures. The Ministry of Infrastructure and Water Management (I&W), however, has established a robust and structured approach through the MIRT model, which aims to improve decision-making, enhance transparency, and ensure high-quality project delivery. This chapter examines the origins, structure, and operational dynamics of the MIRT model, emphasizing its focus on early-stage coordination, cost–benefit analysis, and collaborative planning between national and regional governments.

6.1 Brief Outline and Description of the Context and Policy Basis for the Governance Model

In the Netherlands, there is no joint system for governance of large public investment projects that applies to all ministries. Each ministry has its own processes and procedures for planning, prioritization, and decision-making. The Dutch Ministry of Infrastructure and Water Management (henceforth: I&W) has a well-established project governance system for major infrastructure and water management projects. I&W is responsible for a large portfolio of investment projects in the mobility and transport, the environment, and the water sector. This chapter examines the governance process for such projects.

As far as infrastructure investments are concerned, the Netherlands enjoys special geographical and population advantages over many other countries. It is a small country in terms of land area, but at the same time one of the most densely populated in Europe, with very high GDP/capita (ranked number 4 in Europe in 2023, Statistics Netherlands 2023). Consequently, it is well-placed financially as far as the implementation of infrastructure projects is concerned, as distances are relatively short and expenses can be shared across a

Governance Frameworks for Major Public Projects: International Practices and Experiences, First Edition.
Edited by Vedran Zerjav, Morten Welde and Gro Holst Volden.

large number of people. It is also geographically favoured, inasmuch as most of the country is completely flat. However, it is a major challenge that large land areas are located below sea level and need to be protected against flooding through dykes along the coast and along the numerous rivers and canals. This requires major investments.

A result of the fact that the Netherlands is densely populated is that severe competition exists between governments and stakeholders regarding land use. In many cases, huge debates take place whether a certain location should be used for housing, flood protection, agriculture, nature, transport infrastructure, or energy infrastructure, amongst other things. Although the government aims to combine different types of land use, oftentimes hard choices need to be made. The national government's budget for infrastructure in the Netherlands is around 3% of the overall national government budget each year (Statistics Netherlands 2019).

The Dutch government has three levels: the central government, 12 provinces, and 342 municipalities. Most taxes go to the central government, which then redistributes the revenues to the regional and local levels. The largest share of investments in the national road and rail network and water management projects is funded by the central government. However, a minor share of the costs in most projects is covered through provincial or municipal funds. The co-funding from provinces and municipalities came into way at the time that the energy market was liberalized in the Netherlands, and decentralized governments received a lot of money when selling energy companies to market parties. When the national government did not want to fund a transport project or water management project, the decentral governments used this money to make the national government an offer that they cannot refuse. From that moment, the national government saw co-funding proposals of decentral governments as an indication for the extent to which they really wanted a project. Hence, although co-funding from decentral government is not a formal requirement, it became part of the negotiation. In addition, the national government also looks at decentral governments when further in the planning process the decentral governments ask for additional measures for mitigating nuisance. In terms of co-funding, one exceptional case was the Rijnland Route close to The Hague in which only 50% of the funding came from the central government. Disregarding this exceptional case, in the Netherlands, transport projects are generally best described as centrally provided local public goods as they are predominantly financed from a national, or common pool, tax base. That is, they benefit those living in a particular area or region, while the costs are shared among all taxpayers (e.g. Joanis 2011; Knight 2004).

In the Netherlands, the planning and decision-making process regarding transport infrastructure projects and water management projects is standardized at the national level. A governance system is established called the Multi-year Plan for Infrastructure, Spatial Planning and Transport (MIRT), which is discussed in this chapter. The MIRT model specifies rules and procedures for the development of projects that seek funding from two funds coordinated by I&W: the Mobility Fund and Delta Fund. Only projects that cost more than 112.5 million euros in total in the case of small municipalities and 225 million euros in the case of larger municipalities are subjected to the MIRT model. Municipalities and provinces are expected to coordinate and finance less expensive projects themselves.

6.2 Presentation of the Model

6.2.1 Brief Introduction

The MIRT model was introduced in the 1990s and professionalized around the year 2000 against the backdrop of many negative experiences with public investment projects in the 1990s, such as cost overruns, delays, and benefits that did not materialize. For instance, in the case of the Southern High Speed Railway, De Jong (1999) concluded that underlying studies were not verified for consistency, completeness, clarity, or quality. And with regard to the Betuwelijn – a freight-transport railway line linking the Port of Rotterdam with the European hinterland – it was concluded that the information provided to the House of Representatives to start this project was of poor quality and that it was problematic that the economic effects were estimated using diverse methods (Eijgenraam et al. 2000). In response to this, the central government initialized the Economic Effects Infrastructure Research Program (Onderzoeksprogramma Economische Effecten Infrastructuur, OEEI) to increase methodological consensus and to achieve more systematic project appraisal guidelines. The most important report in the OEEI-program concluded that cost–benefit analysis (CBA) should be the preferred approach for project appraisal (Eijgenraam et al. 2000). Since then, a CBA that is in line with the OEEI Guideline has become compulsory in the decision-making process for all large infrastructure projects. Moreover, the House of Representatives ordered that CBAs for all large national projects should be reviewed in a second opinion to ensure the quality of the CBA (House of Representatives of the Netherlands, 2000).

Furthermore, a special Parliamentary Commission on Infrastructure Projects (TCI, also known as the Duivesteijn Committee) identified many shortcomings in the decision-making process for large infrastructure projects before the year 1998 (Priemus 2007). The Committee concluded that the House of Representatives had been regularly misinformed about the costs and benefits of projects (House of Representatives 2004–2005, 29,283, nrs. 5–6) and emphasized the importance of conducting CBAs for transport infrastructure projects. The official function of CBA was to provide transparent policy information for the preparation of infrastructure projects, and public administrators were not formally bound to the results of the studies. However, ministers who did not act in line with the results had some explaining to do (De Jong and Geerlings 2003).

6.2.2 The Stage-Gate Model with Four Stages

The MIRT model covers the entire process of developing projects and programmes in the physical domain, from the identification of a challenge to the search, finding, and realization of a project. The MIRT model includes guidelines that oversee the organization of joint and phased decision-making. The MIRT guidelines facilitate administrative coordination of (the preparation of) decision-making among the various government parties involved.

The Guidelines state that the structure of the MIRT model is primarily focused on improving the front end of decision-making, from the initial initiative to the implementation decision. Moreover, the aim of the model is to streamline and coordinate various investment initiatives in different parts of the country and to ensure high-quality project proposals. Decision-making on projects without a thorough problem analysis or failure to analyse various alternatives to solve the identified problem should be prevented. Finally, a goal of the structured MIRT model is also to speed up the planning and decision-making process.

The MIRT Guidelines describe the roles and tasks of each government body during a MIRT process. The MIRT Guidelines also contain the national government's decision-making requirements needed to decide on a national financial contribution. For each stage in the MIRT model, the specific information needed for a decision to move a project to the next stage is detailed in an 'information profile'.

A MIRT process consists of four stages:

1) The preparation phase, which may culminate in a start decision.
2) The exploration phase, which may culminate in a preferred alternative decision.
3) The planning and study phase, which may result in a project decision.
4) The construction phase, culminating in a delivery decision.

For each phase, an explicit political decision is taken on whether or not to take the project to the next phase.

Within the Ministry of I&W, the directorates of rail, highways, air, freight, and water are responsible for the project management of the projects that fall within their responsibility. Within the ministry, there is a specific MIRT directorate that coordinates the MIRT. This directorate is responsible for ensuring that the MIRT runs smoothly, but is not responsible for specific projects. The MIRT directorate communicates with the regional actors, looks at the coherence of all the MIRT projects, and provides advice.

6.2.2.1 Preparation Phase

The first stage of the MIRT model is the preparation phase. A preparation phase is initiated by politicians from the national and regional levels. The goal of the preparation phase is that governments mutually identify and investigate challenges and finally – where adequate – outline first solution directions for these challenges. At the end of the preparation phase, at least 75% of the financing of the most obvious solution to the challenge should be in sight to ensure that a project can move to the exploration phase. There is no formal rule to establish what 'the most obvious solution' is. National-level and regional-level policymakers should collaborate to decide this together, basing their decisions on reference projects. The MIRT Guidelines assert that there are no formal requirements for the preparation phase. That is, the MIRT Guidelines prescribe that governments should mutually identify challenges and solutions (and find enough financing), but not how this should be done. When challenges need to be further investigated, the preparation phase can contain a so-called 'MIRT study'. But, when sufficient information is already available about challenges and potential solutions to warrant a decision to (not) start the exploration phase, the preparation phase can be done very quickly, and politicians can directly take a 'start decision', which marks the start of the exploration phase.

The selection of challenges to be investigated in the preparation phase is a political consideration for which no fixed and hard criteria apply (apart from the fact that 75%

financing should be in sight). Various instruments can be deployed to investigate challenges and first solution directions during a MIRT study, ranging from design-based research to running transport models and conducting an indicative CBA.

The preparation of the start decision consists of the preparation of the start document, which begins with a description of the challenge(s). The urgency, nature, and scope of the challenge are described, including substantiation. Also, a description of the development of the challenge(s) without intervention is included. The Guidelines emphasize the importance of taking a broad perspective in describing the challenges to avoid tunnel vision. For instance, for transport projects, the description should not only focus on traffic safety and accessibility but also include other dimensions such as sustainability, energy neutrality, housing, landscape, cultural heritage, knowledge of soil and subsoil, and climate adaptation. No solutions should be taken as the single point of departure of a planning process. Taking a broad perspective means that planners should look beyond improving accessibility and investigate how a programme or project contributes to goals such as quality of life, environmental quality, sustainability and improving international competitiveness.

The start document also describes stakeholders and parties that should be involved, including responsibilities during subsequent phases of the MIRT process. An overview is provided of different 'solutions-pathways' and alternatives that are considered. The start document also describes which funding sources are available to finance potential solutions for the challenge(s). In addition, the governmental agencies involved should investigate whether there are reference projects that are comparable to the most obvious solution for the challenge in terms of the type of solution. If there are reference projects, both the final costs of those projects and the lessons learned are taken into account in the preparation of the cost estimation for the most obvious solution of the challenge(s). There are no formal requirements for how this should be done exactly. An independent cost expert assesses the quality of the cost estimation. Finally, parties identify the main risks that may arise with the most obvious solution.

At the end of the preparation phase, the outputs are discussed during a meeting between politicians from the central government and regional politicians (the 'executive meeting'). There are three possibilities: (1) a start decision is made when a concrete challenge is identified, involved parties are willing to arrange funding for solving the challenge, and feasible solutions are in sight; (2) no start decision is made when the challenge is not pressing enough. In this case, the project is terminated and does not move to the exploration phase; and (3) hold the decision: it is established that solving the challenge(s) may require substantial investment, but that not all conditions for making a MIRT start decision can be met in the foreseeable future (e.g. the challenge is not sufficiently clear or the 75% financing is not in sight yet). Thus, no MIRT exploration can be started yet, but the national and regional governments intend to continue working on solving the identified challenges. In such cases, the decision to start the exploration phase can be put on hold. In this case, the process required to be able to take a start decision should be described and the decision-making is postponed to a future executive meeting. The decisions made during the executive meeting are communicated to the Dutch Parliament including the underlying documentation such as the MIRT study.

6.2.2.2 Exploration Phase

Following a start decision, projects proceed to the MIRT exploration phase in which the aim is to work towards a preferred alternative decision, on the basis of a thorough problem analysis and a broad inventory of solutions. In the exploration phase, the problem analysis and project objectives are concretized; possible solutions are explored; and, eventually, politicians from the national and regional levels make a trade-off between possible solutions. The preferred alternative decision can be either a GO decision, which marks the start of the planning and study phase, or a NO GO decision, which means that the preferred alternative concerns that no solution is implemented. In this case, the planning process is terminated, and the reserved funds are released. A GO decision can only be taken when 100% of the financial resources are available to realize the preferred alternative within the proposed time horizon. Moreover, the MIRT Guidelines prescribe several formal requirements for the tasks conducted in the MIRT exploration phase. For instance, when the challenge is related to transport, it is prescribed to develop and consider at least one non-infrastructural solution to solve the challenge. Moreover, it is required to conduct a CBA, an Environmental Impact Assessment (EIA), and a business case (see Figure 6.1).

Decision-makers can also decide to make an adaptive decision, which means that decisions can already be taken for some parts and not yet for others. This is done by looking at when to take which decision in order to remain sufficiently flexible, on the one hand, but make progress, on the other. The preferred decision may then consist of a package of short-term measures, a concrete preferred alternative, and/or possible long-term solutions that have not yet been fully determined (adaptive), which can be decided on at a later date if required. An agreement may also concern a decision on the timing of the next step in the process, e.g. after the construction of a certain number of houses, in case of an expected rise in sea level by a predefined number of centimetres or actual transport growth by a certain percentage.

A CBA should be conducted in accordance with the Dutch General Guideline for Cost–Benefit Analysis (Romijn and Renes 2013). However, a positive net present value is

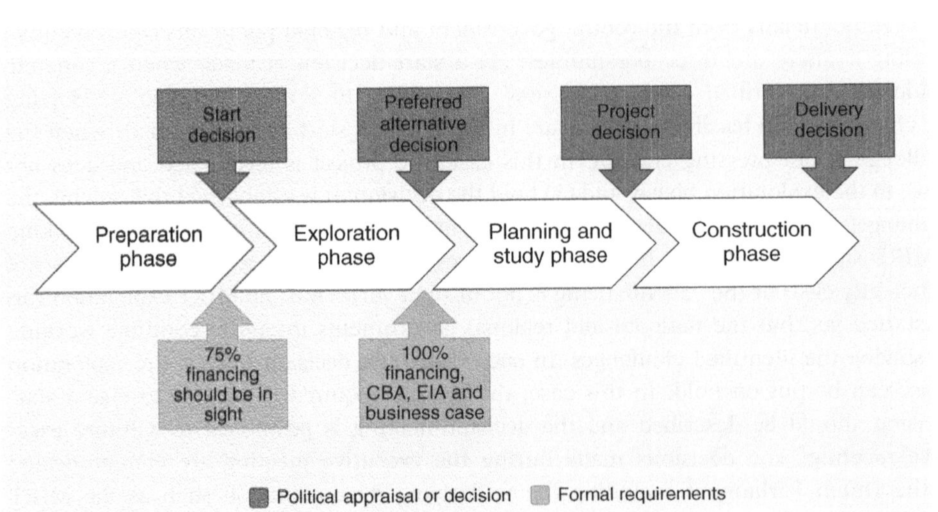

Figure 6.1 The Dutch MIRT model.

not a formal requirement for approved funding (Mouter 2014). The quality of the CBA is also verified by government agencies such as the Netherlands Bureau for Economic Policy Analysis or the Netherlands Institute for Transport Policy Analysis (the verification is hereinafter referred to as the second opinion). All CBAs are published. For decisions requiring an EIA – which applies to almost all physical transport and water projects – an EIA should be conducted during the exploration phase. The EIA is verified by an independent institute: 'the Netherlands Commission for Environmental Assessment'. Both the EIA and the verification should be published. A business case provides information for costing and financing the solution.

Before a preferred alternative decision can be taken, an information profile is drawn up, which should contain a clearly substantiated choice of the preferred alternative, including a substantiation of the rejected solutions. The information profile also includes an overview of the impacts of the preferred alternative and other solutions that have been investigated benchmarked against the reference case, a substantiation of the usefulness and necessity and the costs and benefits of the preferred solution by means of a CBA including the quality control report, insight into the possible risks of the preferred alternative, and a description of the proposed budget including a proposal for the distribution of finances and risks among the involved governments.

When governments decide to take a GO decision, the parties involved lay down all the agreements made in a management agreement. The management agreement contains an elaboration of the administrative, legal, and financial agreements.

6.2.2.3 Planning and Study Phase

The purpose of the planning and study phase is to develop the preferred alternative into a project decision. In this phase, the scope of the project, the realization period, the financing, and the procurement approach are determined. In this phase, the preferred alternative will not change, but optimizations are examined, and expectations regarding uncertainties and possible opportunities will be mapped out.

Before a project decision can be taken, an information profile is drawn up. This information profile contains, amongst other things, the following components: (1) a description of any change(s) or enhancement(s) to the preferred decision; (2) a description of how citizens, businesses, civil society organizations, and administrative bodies will be involved in the implementation of the project; (3) an updated cost estimate. The maximum margin (coefficient of variation) is 15%; (4) an explanation of which form of tendering is preferable and why; (5) a description of the intended approach, lead time, and planning for the construction phase including the necessary licensing and other procedures such as land acquisition; and (6) an administrative-legal paragraph indicating which formal procedure(s) must be completed prior to the construction phase and which formal procedure(s) will be followed during the construction phase.

When a project decision is made, the project proceeds to the construction phase. A project decision is taken if at least the following conditions are met: (1) There is agreement between involved parties on the project alternative to be implemented; (2) Sufficient financial resources are available to realize the project within the proposed plan horizon; (3) Legal and regulatory requirements have been met; and (4) The agreements on funding and financing of the project, accountability, and the risk allocation method are laid down in a management agreement or covenant. At the end of the planning and study phase, it is also possible to take a NO-GO decision. In this case, the planning process is terminated, and the reserved funds are released.

6.2.2.4 Construction Phase

The construction of the project can start after the planning and study phase. In the context of the MIRT model, the progress of the project is monitored during the construction phase. When the project is finished, a 'completion report' is drawn up. This completion report should be finished one year after opening of the project and should contain the following information: (1) financial report of the total project costs; (2) agreements on residual risks and the cost of management, maintenance (and estimate of Life Cycle Costs), and advices for management and maintenance processes; (3) the scope realized and associated major deviations including justifications for delay in project delivery; (4) an assessment of the goals achieved with the project; and (5) recommendations for future projects, for example, in the area of process, tendering, technology, cooperation with contractors, etc.

The completion report provides the basis for a delivery decision, which marks the end of the MIRT process.

6.2.2.5 Portfolio Management

In the MIRT process, the Netherlands is divided into five areas (North-West Netherlands, South-West Netherlands, South Netherlands, East Netherlands, and North Netherlands). For each area, national and regional governments develop a Regional Development Agenda in which governments mutually define the key challenges in an area. The MIRT model has a cyclical annual agenda with several key moments. Every spring, the Minister of Interior Affairs and Kingdom Relations, the Minister of Infrastructure and Water Management, and the Undersecretary of this department visit the five areas. During these 'executive meetings' (BO-MIRT), elected officials at both the national and regional levels discuss the Regional Development Agenda, including ongoing projects and strategic challenges in a specific area. Moreover, they make further financial and administrative agreements where necessary, decide that a project should move to the next planning stage, or decide on studies that are needed to inform such decisions.

The background, status, and planning of ongoing MIRT studies, projects, and programmes are recorded in the annual MIRT Overview. The MIRT Overview is a reference work for the policy context, status, funding obtained, and progress in realizing the agreements made in the executive meetings. The results of the executive meetings and the MIRT Overview are shared with the Parliament annually by the Minister of I&W.

Before the summer, the minister and the undersecretary of I&W discuss the arrangements that are made with the Parliament. In autumn, national-level elected officials and regional-level elected officials again discuss the challenges in the five regions in executive meetings, and in these meetings, investment agreements can be made. At the end of the year, these agreements are discussed in Parliament.

6.3 Experiences

The MIRT model is unique in the Netherlands. Other ministries do not or hardly work with such a model. However, the Ministry of Climate and Green Growth has partly adopted the MIRT model in their Multi-Year Programme Infrastructure and Climate (MIEK). Experiences with the MIRT model are positive. Mainly because a long-term fund allows policy makers to look ahead, it is possible to create a programme that is fixed for a number of years. Without a long-term programme, policymakers will be inclined to focus on the short term. Another strong point is that the MIRT Guidelines give a lot of clarity, certainty, and transparency. As a policymaker, you know what is expected in terms of content and process.

One major change in the MIRT model in 2022 concerned the switch from an 'Infrastructure Fund' to a 'Mobility Fund'. Before 2022, the money was a priori divided percentage-wise among the directorates representing the different modalities. After 2022, there is no priori allocation among modalities and, as a result, different directorates in the Ministry of Transport have to cooperate more. The stronger cooperation between directorates that represent different modalities is experienced in a positive way. There is more of an attempt to look integrally at different modalities when making plans.

Currently, there is quite a lot of discussion in the Netherlands about the rule that 75% of financing should be in place to ensure that a MIRT exploration can be started. Regions in the periphery of the Netherlands are very much against this rule. They feel that this rule stops the planning of important projects. However, from the ministry's perspective, a positive aspect is that the rule ensures that it is possible to manage the amount of work. Without the 75% rule, all kinds of projects are explored for which it is already known a priori that there is not enough money to finance them. The 75% rule ensures that no unrealistic expectations are raised. However, one argument against the 75% rule is that the central government asks a lot from the regions in terms of challenges around building houses, and it is logical that the regions complain about not being able to plan for good public transport access due to the 75% rule.

The MIRT model is more effective for some projects than others. It is particularly suitable for traditional infrastructure projects. But currently, I&W is mainly concerned with maintenance and also with behavioural policies. The MIRT model should probably be adapted for maintenance, replacement, and renovation as intensive alignment with regional actors does not always add that much value. Policies to spread transport across the day and avoid travelling are taken seriously at the I&W, but it is still a lingering question how those policies can have a serious position in the MIRT model.

In recent years, more attention has been paid to quality control and professionalization, especially at the governmental agencies Prorail and Rijkswaterstaat, who are charged with the implementation of transport projects. But due to huge inflation, projects are now facing cost overruns. There is also a political element. Sometimes, in a coalition agreement, a statement is made about a project and how much it will cost, while people know that in reality, the project will cost much more money. But mentioning the actual costs upfront will kill the project. Finally, projects often become more expensive due to additional measures for mitigating noise pollution and encroachment of nature.

6.4 Discussion

The MIRT framework applies to all projects that can possibly be (co-)financed from two funds coordinated by the Dutch Ministry of Infrastructure and Water Management: the

Mobility Fund and Delta Fund. The framework involves four phases, which start with the preparation phase, which is political in nature, and which is followed by the exploration phase, which includes mandatory technical analyses such as a CBA. The MIRT Guidelines are aimed at enabling cooperation between parties based on equality. Central government and decentral government aim to mutually agree on challenges that should be tackled in an area and solutions that should be implemented. During executive meetings between the central government and decentral governments, agreements are made between authorities.

The largest share of investments in transport infrastructure projects is funded by the central government. However, a relatively minor share of the costs is generally covered through provincial or municipal funds, which ensures that local governments, to a certain extent, have 'skin in the game'. Another characteristic of the MIRT model is that already at an early stage of the planning process, 75% of financing should be in sight. This rule was established to prevent investment in the planning process of projects for which it is already known a priori that there is not enough money to finance them. However, at the moment, a discussion rages in the Netherlands whether this rule is (not) too restrictive.

In the Dutch governance process, the quality of some studies is checked by (independent) experts. For instance, an independent cost expert assesses the quality of the cost estimation, the quality of the CBA is verified by economists from government agencies, and the EIA is assessed by an independent committee. This is different from, for instance, the Norwegian governance process in which a whole planning phase is reviewed by independent experts. Despite the attention that has been paid to quality control and professionalization, projects currently face cost overruns due to huge inflation.

In the Netherlands, there is no systematic ex post evaluation of transport projects. The formal governance process ends one year after the opening of the project. This contrasts, for instance, the Norwegian governance process in which some projects are evaluated annually based on a standardized evaluation model that includes six criteria covering operational, tactical, and strategic success (Volden and Welde 2022).

References

De Jong, W.M. (1999). *Institutional Transplantation; How to adopt Good Transport Infrastructure Decision-Making Ideas from other Countries?* Delft: Eburon Publishers.

De Jong, W.M. and Geerlings, H. (2003). Exposing weaknesses in interactive planning: the remarkable return of comprehensive policy analysis in The Netherlands. *Impact Assessment and Project Appraisal* 21 (4): 281–291.

Eijgenraam, C.J.J., Koopmans, C.C., Tang, P.J.G., and Verster, A.C.P. (2000). Evaluation of Infrastructural Projects; Guide for cost-benefit analysis, Sections I and II, CPB, The Hague, NEI (Changed Name to ECORYS), Rotterdam.

House of Representatives (2000). Letter of Minister of Transport, Public Works and Watermanagement 26428 nr. 21 (Perspectievennota Verkeer en Vervoer). The Hague, April 20, 2000.

House of Representatives (2004). Grote Projecten Uitvergroot: Een Infrastructuur voor Besluitvorming (The Hague: Tweede Kamer der Staten-Generaal, 2004, TK 2004–2005, 29 283, nr.5-6).

Joanis, M. (2011). The road to power: partisan loyalty and the centralized provision of local infrastructure. *Public Choice* 146 (1–2): 117–143.

Knight, B. (2004). Parochial interests and the centralized provision of local public goods: evidence from congressional voting on transportation projects. *Journal of Public Economics* 88 (3–4): 845–866.

Mouter, N. (2014). Cost-benefit analysis in practice. A study of the way Cost-Benefit Analysis is perceived by key individuals in the Dutch CBA practice for spatial-infrastructure projects. Delft.

Priemus, H. (2007). Decision-making on large infrastructure projects: the role of the Dutch Parliament. *Transportation Planning and Technology* 30: 71–93.

Romijn, G. and Renes, G. (2013). *Dutch General CBA Guideline*. Algemene Leidraad voor maatschappelijke kosten-batenanalyse. Centraal Planbureau en Planbureau voor de Leefomgeving (In Dutch).

Statistics Netherlands (2019). Spendings of the Dutch national government.

Statistics Netherlands (2023). Dutch GDP per capita in 2023 ranks fourth in EU. https://www.cbs.nl/en-gb/news/2024/17/dutch-gdp-per-capita-in-2023-ranks-fourth-in-the-eu.

Volden, G.H. and Welde, M. (2022). Public project success? Measuring the nuances of success through ex post evaluation. *International Journal of Project Management* 40: 703–714.

7

Denmark

Per Svejvig[1] and Andreas Claus Hansen[2]

[1]*Department of Management, Aarhus University, Aarhus, Denmark*
[2]*Danish Building and Property Agency, Copenhagen, Denmark*

Denmark has developed two distinct governance models to manage large-scale public investments: one for transport infrastructure and another for state-funded IT projects. These models emerged in response to past challenges, including cost overruns and project delays, and now serve as structured frameworks to improve decision-making, budgeting, and risk management. While not universally applied across all sectors, they have become influential in shaping governance practices for major public projects. This chapter outlines the origins, structure, and operational mechanisms of Denmark's infrastructure and IT project models, highlighting their role in enhancing transparency, accountability, and delivery performance in public investment

7.1 Context and Policy Basis

Denmark, with a population of 5.9 million, ranks among the top 15 countries by GDP per capita (World Bank 2024) and first in the 2023 Transparency International Corruption Perceptions Index (Transparency International 2023). The economy is robust, driven by strong domestic demand and exports, with low unemployment and rising labour force participation (European Commission 2024). Public finances are sound, with a budget surplus since 2017 and low public debt (OECD 2024b). Denmark is committed to its green transition, making significant strides in renewable energy and sustainability (European Commission 2024).

Denmark operates with a three-tier government system, where the responsibilities of the central government, regions, and municipalities are distinctly outlined. The central government is responsible for national policies, legislation, and governance. The five regions primarily handle healthcare services, regional development, transport, and environmental planning. Municipalities, of which there are 98, manage a wide range of local services, including social welfare, education, local roads, and public utilities (Council of Europe 2013).

Governance Frameworks for Major Public Projects: International Practices and Experiences, First Edition.
Edited by Vedran Zerjav, Morten Welde and Gro Holst Volden.

Denmark has a high standard of living supported by well-designed policies. Generous social support and public spending are financed through comparatively high taxes. A pension reform has helped to improve financial robustness (OECD 2024a). Denmark substantially depends on international trade and a few firms and sectors, such as the pharmaceutical industry (OECD 2024b).

Denmark shares the challenges of other Western countries, such as the ageing population and labour shortages. As a consequence of digital and green transitions and the increasing demand for long-term care, longer working lives must be encouraged, and persistent labour shortages must be addressed (OECD 2024a,b).

Denmark is strongly committed to the green transition and has implemented ambitious targets for reducing greenhouse gas emissions. Accelerating electrification and expanding green power production capacity are required to move to a carbon-neutral economy (OECD 2024b), which demands extensive infrastructure investments and further increases the need for labour.

Public investments in Denmark are substantial and estimated to amount to approximately DKK 70 billion in 2024 (≈€9.4 billion) and 2.5% of GDP. The investments are expected to grow due to increased defence acquisitions and the focus on the green transition (Økonomiministeriet 2024). Denmark excels in digital infrastructure, ranking second in government digitalization indices, with strong internet access and digital public services (OECD 2023). While the country performs well in road and air transport infrastructure, there is room for improvement in railway infrastructure. Both physical and digital infrastructures are crucial for Denmark's competitiveness, productivity, and economic growth. Enhancing these areas, focusing on artificial intelligence and innovative technologies, is vital for future economic and societal development (Dansk Erhverv 2024).

Danish workplace culture features a flat hierarchy, valuing teamwork and individual responsibility. The standard workweek is 37 hours, with a strong emphasis on punctuality and deadlines. Employees enjoy independence and flexible working hours. Proactivity and informal communication are encouraged, with everyone addressed by their first name. Titles and status are downplayed, fostering a humble and collaborative environment (Work in Denmark 2024).

7.2 The Two State Project Models

Denmark has two state project models, one for roads and railways, which has also inspired the governance of other types of construction projects (Transportministeriet 2024a), and one for IT projects (Økonomistyrelsen 2024d). They are both subject to budget guidance from the Ministry of Finance (Finansministeriet 2021a). However, neither of the two models covers several areas, including energy infrastructure and defence acquisitions.

Ministries, agencies, and state-owned enterprises may have their own project models, based on either the overall state project models or entirely their own, depending on regulations for the given area. Similarly, local governments, such as municipalities and regions, have their own governance frameworks for locally funded projects.

The two state project models are described in more detail in the subsequent sections.

7.2.1 A Brief Historical Overview

7.2.1.1 The New Infrastructure Budgeting

Around the millennium, several large-scale public investment projects experienced project delays and cost overruns (Transport og Energiministeriet 2006), attracting both media and political attention. Among these are the Great Belt Link and the concert hall for the Danish Broadcasting Corporation.

In surrounding countries, such as Norway and the United Kingdom, initiatives were already implemented to strengthen project governance, like the first version of the Norwegian project governance scheme preluding the State project model (the Norwegian chapter of this book) and the Green Book from HM Treasury in the United Kingdom (HM Treasury 2022). In its approach to project governance, Denmark followed the lead of the United Kingdom (Flyvbjerg 2014).

The Danish Ministry of Finance analysed the status of major investment projects and made recommendations for an overall project governance scheme, primarily for public infrastructure projects.

In 2003, it was decided to structure public investment decisions in two levels: (1) Level 1, the decision on drafting a detailed basis for investment decision and (2) Level 2, the final investment decision on the project (Transport og Energiministeriet 2006).

New principles for budgeting infrastructure projects under the Ministry of Transport were implemented in 2007 (Transport og Energiministeriet 2006). These principles were formalized under the headline of 'New Principles for Infrastructure construction budgeting', and the principles also influence the cost and risk management of the projects that must adhere to the new principles (Transportministeriet 2024a).

The principles included two additional instruments: (1) mandatory external quality assurance (QA) and (2) experience-based budget uplifts. The purpose of the principles was to strengthen the decision basis of both decision levels 1 and 2 (Transport og Energiministeriet 2006).

With the original intent of securing proper budgeting for infrastructure construction projects, the scheme has widened over the years. It is still referred to as a budget scheme, even when the primary content relates to risk management and QA. In recent years, these principles have also been applied to building construction projects in addition to infrastructure, and as such, the principles act as a frame of reference for large-scale project governance (Transportministeriet 2021).

Still, there are no official initiatives to harmonize practices across different line ministries or areas of investment. As such, it does not hold a formal position as a requirement for entities outside the Ministry of Transport.

7.2.1.2 The State IT Project Model

Denmark experienced major challenges with IT projects through the 1990s and 2000s. A report from 2001 stated that the main challenges with IT projects in Denmark were weak leadership and strategic focus, lack of organizational readiness, poor project management practices, and insufficient inter-organizational coordination (Teknologirådet 2001). The problems continued throughout the 2000s, but in 2009, a working group was set up by the government, which made the following recommendations: (1) adopt a standardized project model for all state IT projects, (2) enhance competence development through certification and training, and (3) focus on high-risk projects with systematic risk assessments (Finansministeriet 2010) as well as establishing a new state IT council (Finansministeriet 2021b). It also suggested to (4) improve collaboration with suppliers through continuous evaluation and dialogue and (5) establish a central project office to oversee project models and competence development initiatives (Finansministeriet 2010).

The Danish State IT Project Model was thus introduced at the very beginning of 2011 (Økonomistyrelsen 2011) and has included many updates since then. The main elements include a structured phase model for project delivery, comprehensive management documents and guidelines, mandatory risk assessments for projects over 15 million DKK, and regular status reporting to the State IT Council. The model aims to ensure successful project delivery within a common management framework, supporting state authorities in achieving their IT project ambitions. It helps manage risks, supports flexible project processes, and encourages innovative approaches while avoiding rigid constraints (Økonomistyrelsen 2024b).

There are signs that the initiatives from 2010 to 2011 have remedied various issues with IT projects (Alami et al. 2021, 2022; Krancher et al. 2022). However, a recent report shows that there are still significant challenges, especially for the very large IT projects that cost more than 300 million DKK (Finansministeriet 2023). The following measures have been initiated: strengthened management of very large IT projects exceeding 300 million DKK, enhancement of the State IT Council's advisory efforts, improved cross-sectional overview of IT project expenses, and the establishment of a uniform conceptual framework for recording IT expenses (Økonomistyrelsen 2024c).

7.2.2 The Stage-Gate Model with Phases and Decision Points

7.2.2.1 The New Infrastructure Budgeting Model

Figure 7.1 presents the New Infrastructure Budgeting Model. This model applies to all road and rail construction projects approved through a project-specific construction act or that is included in the national budget law (annual finance act) with its own grant (Transportministeriet 2024a).

All these projects are under the purview of the Ministry of Transport, primarily in the Danish Road Directorate and Rail Net Denmark. Projects above a threshold of 350 million DKK, including a mandatory budget uplift, have a mandatory requirement for external QA. The phases and phase names may vary depending on the type of project, but the principles

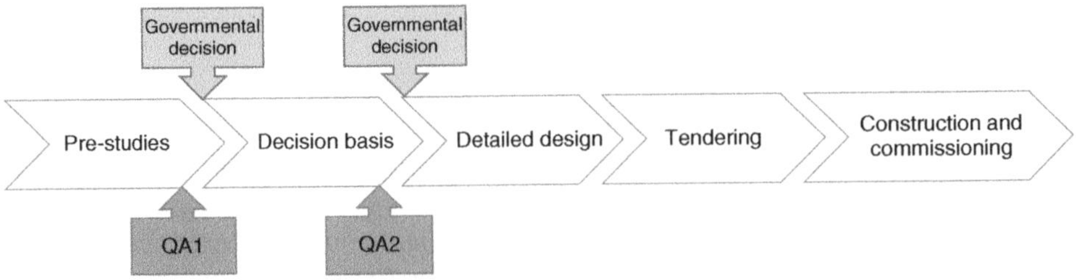

Figure 7.1 New infrastructure budgeting model under the Ministry of Transport. *Source:* Adapted from Transportministeriet (2024a).

remain the same, as well as the timing of external QA (Transportministeriet 2024a). There are many similarities with the Norwegian State Project Model.

The phases are described herein.

Pre-studies: The pre-studies phase involves the initial evaluation and analysis to determine the problem a project aims to solve and whether it would have a positive economic impact. This phase is the very front-end phase, with few strict requirements on the part of the investigating entity. There can be pre-studies for many parallel projects, but only projects that pass approval at the political level will proceed to the next stage. If, based on the pre-studies, a project is deemed relevant to proceed with, the first major decision gate must be passed. For major projects, this entails the passing of a law approving the design of the project, e.g. the design law for the Femern Belt Link project (Transportministeriet 2009).

Decision basis: This phase concludes all the investigations necessary to have a solid basis for decision-making on whether to carry out the project. The primary deliverable is an environmental assessment covering various environmental topics. After completion, the final investment decision can be made to approve the delivery of the project. For major projects, this entails the passing of a law approving the final design and construction of the project, e.g. the construction law for the Femern Belt Link project (Transportministeriet 2015).

Detailed design: Final design of the project, preparing for tendering and delivery of the actual construction works.

Tendering: Tendering process for choosing contractors for the execution of construction works.

Construction and commissioning: Delivery of the construction works on sites and taking the infrastructure into operation etc.

The last three phases are not sharply divided, as different parts of the project may be in different stages. For these, there is a requirement to report on the progress (Transportministeriet 2024a). Progress reporting is made public in bi-annual status reports by the Ministry of Transport (Transportministeriet 2024b).

7.2.2.2 The Danish State IT Project Model

Figure 7.2 presents the current Danish State IT Project Model. The model applies from the project idea is conceived, analysis including funding decision, delivery with the launch, and, finally, realization (Økonomistyrelsen 2024b).

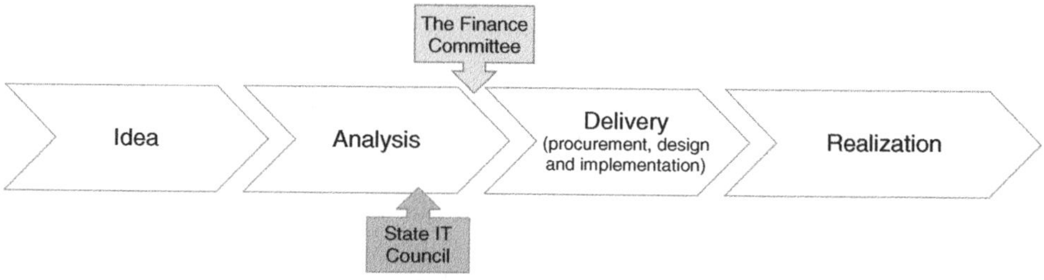

Figure 7.2 The Danish State IT Project Model.

The phases are described herein.

Idea phase: This is the initial phase where the project idea is qualified and significant benefits are identified. The purpose is to determine if the idea is viable, fits within the institution's IT portfolio, and if resources are available for analysis and delivery. Preliminary risk analysis and stakeholder mapping are also conducted.

Analysis phase: In this phase, the project is detailed, and a management framework is established, including an extended project description, risk checklist, risk log, business case model, steering group agreement, and project overview. If the budget exceeds 15 million DKK and 70 million DKK for IT acquisition projects, the State IT Council assesses the project's risk. If the budget exceeds 70 million DKK, the Danish Parliament Finance Committee must approve the project, cf. the budget guidelines (Finansministeriet 2021a).

Delivery phase: This phase involves the procurement, design, and implementation of the project's deliverables. It includes requirements specification, procurement, development of deliverables, and organizational and technical implementation. Parts of the benefit realization may already be underway, and the project basis and business case are continuously updated.

Realization phase: At this stage, the project is completed, and the realization of benefits is handed over to operational units. The focus is to ensure that the benefits described in the business case are achieved and integrated into the organization. This phase does not have a formal end and continues until the benefits are fully realized. About one year after the completion of the project, a report on the realization of benefits must be submitted.

State IT projects with total budgeted expenses of DKK 15 million or more must use the State IT Project Model. If the project's total budgeted expenses amount to less than 15 million DKK but more than one million DKK, the projects are required to establish a governance structure similar to the project governance structure in the State IT Project Model but in a simplified version according to the project size. IT projects whose total budgeted expenses are estimated to amount to less than DKK 15 million, but which later develop to amount to DKK 15 million or more must use the model and be risk assessed by the State IT Council. The same applies to acquisitions of standard IT solutions, which amount to DKK 70 million or more (Økonomistyrelsen 2024b).

7.2.3 Projects That Must Adhere to the Model

The New Infrastructure Budgeting Model's scope includes road and rail construction projects but is also used in building construction projects. No minimum size has been specified for projects that follow the new construction budgeting. The principles apply to all relevant construction projects, regardless of size.

The State IT Project Model should be used for all government-funded IT projects with the above-mentioned threshold values. This applies to all ministries, agencies, and state-owned institutions.

The scope for the two models is well defined, but this also means several areas are not covered, including areas with very high investment. Defence projects are not included; they follow their own project model with the phases of idea, planning, procurement, contract, implementation, and completion (Deloitte 2024; Rigsrevisionen 2024). The Ministry of

Defense provides funding authority to carry out the procurement, cf. the budget guidelines (Finansministeriet 2021a), similar to the State IT Project Model. DKK 190 billion must be invested in defence in the period 2024–2033 (Forsvarsministeriet 2024), so these are substantial and often risky investments, and the entire defence sector has been under considerable criticism for their lack of ability to implement projects (Deloitte 2024; Rigsrevisionen 2024). We see a parallel pattern to the defence area for climate, energy, and the utility area, which are also not covered by the two national models.

Another related issue is which model to choose when you have two national models. The Signal Programme must modernize and digitize the Danish railway network to improve safety and efficiency, and it costs approx. 22 billion DKK. The project is a railway construction project and an extensive IT project. The project follows the New Infrastructure Budgeting Model, but it is primarily the IT area that has caused problems in the project (Banedanmark 2024). Perhaps the project could have benefited from using the State IT Project Model, which also involves advice from the State IT Council.

7.2.4 Parties and Roles

7.2.4.1 The New Infrastructure Budgeting Model

The organization of the scheme is illustrated in Figure 7.3 and described as follows:

- *The Agency for Public Finance and Management*, under the Ministry of Finance, is responsible for the budget guidance documents, which are the comprehensive management documents and guidelines for all public investments
- *The Ministry of Transport* is responsible for the New Infrastructure Budgeting model and the relevant documents, as required by the regulation in the law on infrastructure budgeting.
- *Government agencies reporting to the Ministry of Transport* are responsible for setting up processes that adhere to the principles in the model, collecting data,

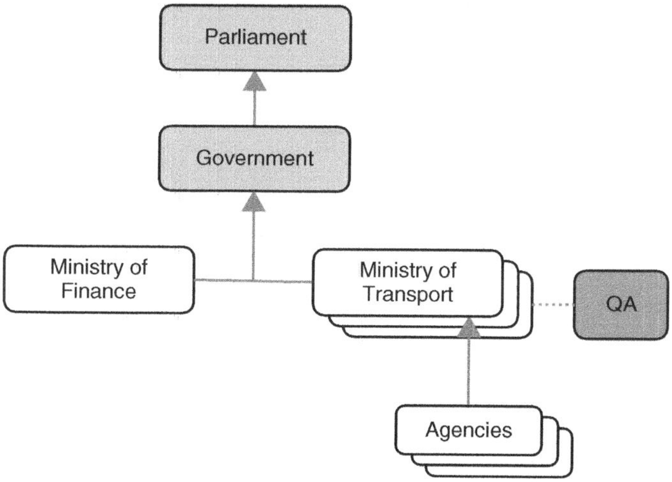

Figure 7.3 The New Infrastructure Budgeting governance.

analysing it, and structuring applications for project funding in accordance with the principles and requirements.

- For projects above the threshold of DKK 350 million, the external QA is carried out by external consultants based on standard ‘terms of reference’.
- Depending on the type of project funding, there can be a requirement for the Ministry of Transport and the responsible agency to prepare a document for approval by the parliamentary *Finance Committee.*

7.2.4.2 The State IT Project Model

The organization of the **State IT Project Model** is illustrated in Figure 7.4 and described as follows:

- *The Agency for Public Finance and Management* is responsible for the State IT Project Model, including the phase model for project delivery, comprehensive management documents, and guidelines (Økonomistyrelsen 2024d).
- The *line ministries* and *government agencies* are responsible for managing and controlling the four project phases: idea, analysis, delivery, and realization. To ensure that decisions, management, and governance are anchored in the right places, the IT project model includes three levels of governance (Økonomistyrelsen 2024b).
- Risk assessment and general advice are a central part of project governance, and the *State IT Council* handles this. All IT projects above a threshold must submit a project overview and description, risk log and checklist, business case, and steering group agreement to the first round of meetings with the council. The council makes several recommendations for

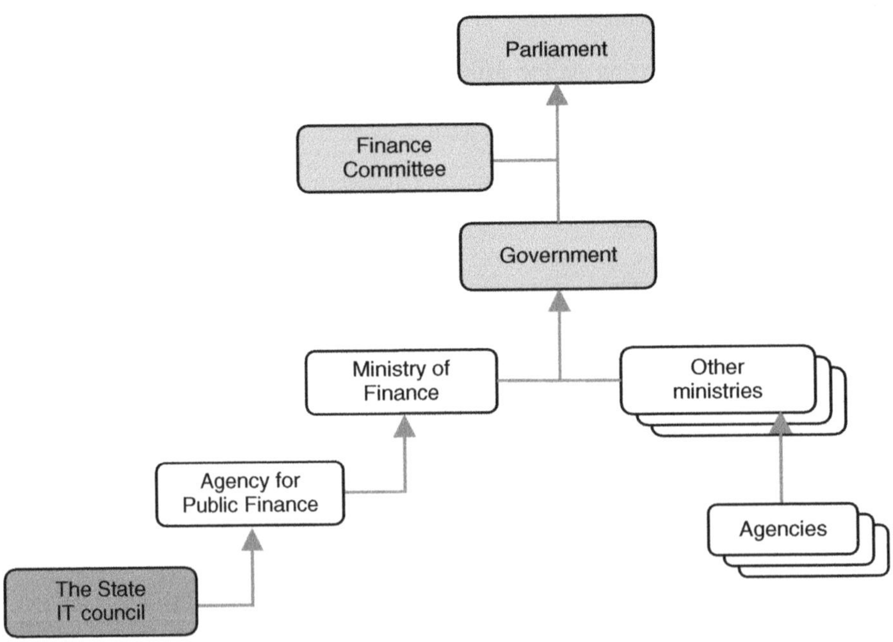

Figure 7.4 State IT Project Model governance.

the project, which must be addressed before the project is approved at the parliamentary level. The council then follows the project by submitting half-yearly status reports. At the end of the project, a final status report must be submitted, and a benefit realization report one year after completion (Økonomistyrelsen 2024a).

- If a government project has a budget exceeding DKK 70 million, a parliamentary document is prepared by the Line Ministry and for agencies in collaboration with the ministry to which they belong. This should then be approved by the Parliamentary *Finance Committee.*

7.2.5 Selected Topics

7.2.5.1 Cost Estimation and Budget Management

The New Infrastructure Budgeting was introduced to remediate cost overruns in public transport infrastructure projects (Transport og Energiministeriet 2006). In the analysis leading up to the implementation of the model, there was a focus on making project budget estimates more robust and accounting for the uncertainty of estimating on different levels of information as the projects are planned (Trafikministeriet 2003). The model, therefore, introduced an experience-based budget uplift of 50% at the first decision gate to improve the basis for prioritizing between projects and 30% at the second decision gate, which were included in the project financing as a contingency for the project's execution (Transport og Energiministeriet 2006). One-third of the fixed contingency is allocated for the project and is governed by the agency responsible for the project. Two-thirds is a central contingency governed by the Ministry of Transport, which acts as a portfolio reserve (Transport og Energiministeriet 2006).

Cost estimation and budget management are crucial components in the State IT Project Model. The model emphasizes creating a detailed project basis, including a comprehensive budget plan. This plan outlines all expected costs and is continuously updated throughout the project life cycle. Regular risk assessments are conducted to identify potential financial risks, mitigate them effectively, and include a risk contingency in the budget. The model also requires periodic status reports to ensure the project remains within budget. A business case is also developed to justify the investment, detailing the expected benefits and costs. This structured approach helps maintain financial control and ensures the project's economic viability (Økonomistyrelsen 2024b).

7.2.5.2 Quality Assurance

The New Infrastructure Budgeting Model applies external QA to strengthen the quality of the decision basis, improve the political management of cost, and prioritize large-scale infrastructure projects. This is done by providing an independent review of the project material and budget, including a review of the budget estimate, traffic and technical solution, project organization, and the analysis of the return on investment. The assurance should qualify the project material and provide an account of whether there are important reasons not to approve the project. The external review does not include investigating alternatives or providing other solutions, but only reviews the provided project material (Transportministeriet 2022a). For level 2, the assessment should be considerably more detailed and thorough, as that forms the basis for the final political investment decision (Transportministeriet 2022b).

The State IT Council oversees and ensures the quality of IT projects within the public sector. They guide and support government agencies to help them manage IT projects effectively. The council conducts regular reviews and assessments of ongoing projects to identify potential risks and issues early on. They also establish the best practices and standards for project management, ensuring that projects are delivered on time, within budget, and meet the required quality standards. Additionally, the council promotes the use of modern technologies and methodologies to improve efficiency and effectiveness in public sector IT projects (Økonomistyrelsen 2024a). 'Both project costs and operating costs must be estimated via a recognized estimation method, which takes the height of uncertainty, and which can reduce the uncertainty of an estimate' (Økonomistyrelsen 2024d, p. 8), but there is no requirement to use stochastic methods, reference class forecasting, etc.

7.2.5.3 Reporting of the Project Portfolio

Denmark is relatively open about the project portfolio status for the two models. For both models, status reports are published every six months, and several key performance indicators are followed up.

The status report for projects following the New Infrastructure Budgeting Model is used to monitor and evaluate the progress of ongoing projects and initiatives. It serves as a tool for transparency and accountability, ensuring that stakeholders are informed about the status, challenges, and achievements. The report includes key performance indicators (KPIs) such as project milestones, budget adherence, use of contingency, and timelines (Transportministeriet 2024b).

The status reports from the State IT Council are used to evaluate the progress and management of government-funded IT projects. It provides an overview of the status of critical IT systems, project portfolios, and IT costs. The report highlights key challenges, such as balancing activity levels with available capacity and ensuring projects are completed on time and within budget. Performance indicators tracked include project expenses, adherence to timelines, and the realization of economic and non-economic benefits. The accompanying annex details the status of individual IT projects, using a traffic light system (green, yellow, and red) to indicate their progress (Statens IT-Råd 2024).

7.3 Experience with the Two Models

In 2019, the national audit agency (Rigsrevisionen) reviewed 32 completed road infrastructure projects under this scheme. It concluded that 30 came in under budget, with a total cost underrun of DKK 11.6 billion. The state auditors concluded that the previous problems of cost overruns had been resolved but that overbudgeting is not satisfactory either, as it binds public financial assets that could have been used for other investments (Rigsrevisionen 2019). The review recommended revising the contingency allocations, and as a result, the mandatory budget uplift was halved for ordinary road projects, but not for other project types, as these have not shown the same trend (Rigsrevisionen 2022). In the latest report from the national audit agency on state-governed building and infrastructure construction projects, published in 2025, a total of 120 projects, with a combined total budget of DKK 91 billion were reviewed (Rigsrevisionen 2025). The report looked into projects managed by the six different ministries having the largest expenditure for building and infrastructure construction projects, and thus, it covered projects related to construction of roads, railroads, university facilities, barracks and prisons, etc. It concludes that 60% of the projects are delayed and 20% experience cost overruns, while 40% are completed with at cost underrun. It further concludes that building construction projects are more often delayed than infrastructure construction projects and that cost underruns are more common for infrastructure construction projects with 56% finishing below initial budget while the same is only true for 32% of the building construction projects (Rigsrevisionen 2025). The report recommends that the Ministry of Finance looks into how to generally improve the decision basis in relation to both schedule and budget before parliamentary project approval (Rigsrevisionen 2025).

A collection of studies investigated various IT projects (Alami et al. 2021, 2022; Krancher et al. 2022). These studies showed that 54 IT projects reported a minor mean cost overrun of less than 10% and a maximum of 170%. Schedule overruns were more pronounced, with a mean of 31% and a maximum of 214%. The IT projects had a mean of DKK 40 million and a maximum of DKK 139 million (Alami et al. 2022). The authors concluded that the cost and schedule overruns were significantly lower than in earlier studies (Alami et al. 2022). A report from the Ministry of Finance highlights that larger IT projects over DKK 300 million do not necessarily have higher proportional cost overruns compared to smaller projects. However, the absolute financial impact of overruns on these large projects is significant. For instance, six projects with absolute overruns collectively accounted for DKK 880 million in additional costs. This suggests that while the percentage of overrun might not be higher, the financial risk and consequences are substantial due to the high baseline costs of these projects. The report emphasizes the need for stringent cost management and risk assessment for these high-value projects to mitigate potential financial impacts (Finansministeriet 2023).

Acknowledgements

Many thanks to the Danish Agency for Public Finance and Management and the Ministry of Transport for reviewing this book chapter and making constructive suggestions for improvements related to the description of each of their respective project models. However, the authors are fully responsible for the content of the book chapter.

Microsoft Copilot has been used comprehensively for literature searches, text (re)formulations, language translations, and document analysis, such as creating summaries.

References

Alami, A., Madsen, C., and Krancher, O. (2021). Spotlight on the positives: how do information technology projects achieve cost underruns? *Hawaii International Conference On System Sciences 2021 (HICSS-54).*

Alami, A., Østergaard Madsen, C., and Krancher, O. (2022). *Government IT Projects: Current Evidence of Cost and Schedule Overrun and Their Antecedents*. Cham: Springer International Publishing.

Banedanmark (2024). Signalprogrammet Statusrapport juli-august 2024 [Signal programme status report July–August 2024].

Council of Europe (2013). *Local and Regional Democracy in Denmark*. Strassbourg: Council of Europe.

Dansk Erhverv (2024). Konkurrencebarometer for transport og digitalisering [Competition barometer for transport and digitalisation]. https://www.danskerhverv.dk/rammevilkar/transport-og-digitalisering/ (accessed 27 August 2024).

Deloitte (2024). Analyse af Forsvarsministeriets Materiel- og Indkøbsstyrelses (FMI) organisation, processer og ressourcer [Analysis of the Ministry of Defence's Material and Procurement Agency (FMI) organization, processes and resources].

European Commission (2024). 2024 country report – Denmark, Commision staff working document. Brussels: European Commision.

Finansministeriet (2010). Professionalisering af arbejdet med it-projekter i staten [Professionalisation of the work with IT projects in the state].

Finansministeriet (2021a). Budgetvejledning [Budget Guidance]. København: Finansministeriet.

Finansministeriet (2021b). Cirkulære om Statens It-råd [Circular about the State IT Council]. København: Finansministeriet. **CIR1H nr 9254 af 21/04/2021**.

Finansministeriet (2023). Budgetredegørelse for udgiftspolitisk styring af it-området [Budget statement for expenditure policy management of the IT area]. København: Finansministeriet.

Flyvbjerg, B. (2014). What you should know about megaprojects and why: an overview. *Project Management Journal* 45 (2): 6–19.

Forsvarsministeriet (2024). Bredt flertal bag fremrykning af investeringer i Forsvaret [Broad majority behind advancing investments in the Armed Forces]. https://www.fmn.dk/da/nyheder/2024/bredt-flertal-bag-fremrykning-af-investeringer-i-forsvaret/ (accessed 15 January 2025).

HM Treasury (2022). *The Green Book - Central Government Guidance on Appraisal and Evaluation*. London: HM Treasury.

Krancher, O., Christian, Ø.M., Alami, A., and Petersson, C.P. (2022). Explanations for budget and schedule overrun revisited – a configurational perspective on IT projects. *International Conference on Information Systems (ICIS) 2022 Copenhagen*.

OECD (2023). Digital government index: 2023 results.

OECD (2024a). Denmark: reforms needed to tackle labour shortages, adapt to population ageing and achieve the green transition. https://www.oecd.org/en/about/news/press-releases/2024/01/denmark-reforms-needed-to-tackle-labour-shortages-adapt-to-population-ageing-and-achieve-the-green-transition.html (accessed 26 August 2024).

OECD (2024b). *OECD Economic Surveys: Denmark 2024*. Paris: OECD Publishing.

Økonomiministeriet (2024). Økonomisk redegørelse maj 2024 [Financial statement May 2024]. København K: Økonomiministeriet

Økonomistyrelsen (2011). Vejledning om den fællesstatslige statens it- projektmodel v1.0 [Guidance on the federal government's IT project model v1.0]. Økonomistyrelsen.

Økonomistyrelsen (2024a). It projekter: Vejledning til risikovurdering og rådgivning ved Statens It råd [IT projects: Guidance for risk assessment and advice at the Norwegian IT Council]. København: Økonomistyrelsen.

Økonomistyrelsen (2024b). Statens it-projektmodel [The state's IT project model]. https://oes.dk/it-og-oekonomistyring/it-projektstyring/statens-it-projektmodel/ (accessed 29 August 2024).

Økonomistyrelsen (2024c). Tiltag til opfølgning på Finansministeriets budgetredegørelse for udgiftspolitisk styring af it-området [Initiatives for follow-up on the Ministry of Finance's budget statement for expenditure policy management of the IT area]. https://oes.dk/it-og-oekonomistyring/budgetredegoerelse-for-udgiftspolitisk-styring-af-it-omraadet/ (accessed 29 August 2024).

Økonomistyrelsen (2024d). Vejledning til statens it-projektmodel [Guide to the state's IT project model]. Økonomistyrelsen.

Rigsrevisionen (2019). Beretning afgivet til Folketinget med Statsrevisorernes bemærkninger -Budgetteringen af Vejdirektoratets vejprojekter [Report submitted to the folketing with the State Auditors' Comments - The Danish Road Directorate's budgeting of road projects]. Rigsrevisionen.

Rigsrevisionen (2022). Notat om beretning om budgetteringen af Vejdirektoratets vejprojekter [Note on the report on the budgeting of the Road Directorate's road project]. Rigsrevisionen.

Rigsrevisionen (2024). Beretning afgivet til Folketinget med Statsrevisorernes bemærkninger -Forsvarsministeriets større anskaffelser af materiel [Report submitted to the folketing with the State Auditors' Comments - The Ministry of Defence's major acquisitions of equipment]. Rigsrevisionen.

Rigsrevisionen (2025). Tidsforbrug og økonomi i statslige bygge- og anlægsprojekter [Time management and budgeting in government construction and infrastructure projects]. Rigsrevisionen.

Statens IT-Råd (2024). Statusrapport 2023 [Status report 2023]. København: Statens IT-Råd.

Teknologirådet (2001). Erfaringer fra statslige IT-projekter - hvordan gør man det bedre?. Rapport og anbefalinger fra en arbejdsgruppe under Teknologirådet. København: Teknologirådet.

Trafikministeriet (2003). *Anlægsprojekter på trafikområdet – planlægning og beslutning*.

Transparency International (2023). Corruption perceptions index 2023. Berlin: Transparency International.

Transport og Energiministeriet (2006). Aktstykke om nye budgetteringsprincipper [Act on new principles for budgeting]. Transport og Energiministeriet. Copenhagen. **Aktstykke nr. 16, Finansudvalget, Folketinget**.

Transportministeriet (2009). Lov om projektering af fast forbindelse over Femern Bælt med tilhørende landanlæg i Danmark [Act on the planning of a fixed connection across the Fehmarn Belt with associated land facilities in Denmark]. Transport **LOV nr 285 af 15/04/2009**.

Transportministeriet (2015). Lov om ændring af lov om Sund og Bælt Holding A/S, lov om projektering af fast forbindelse over Femern Bælt med tilhørende landanlæg i Danmark og selskabsskatteloven [Act amending the Act on Sund og Bælt Holding A/S, Act on the design of a fixed connection across the Fehmarn Belt with associated land facilities in Denmark and the Corporation Tax Act]. Transport Copenhagen, Transportministeriet. **LOV nr 581 af 04/05/2015**.

Transportministeriet (2021). Aktstykke om Odense kontorknudepunkt [Document about the Odense office hub]. Transportministeriet. Copenhagen. **Aktstykke nr. 69, Folketinget 2021–22.**

Transportministeriet (2022a). Terms of reference - Ekstern kvalitetssikring af fase 1 undersøgelser [Terms of reference - External quality assurance of phase 1 studies]. København: Transportministeriet.

Transportministeriet (2022b). Terms of reference - Ekstern kvalitetssikring af fase 2 undersøgelser [Terms of reference - External quality assurance of phase 2 studies]. København: Transportministeriet.

Transportministeriet (2024a). Hovednotatet for Ny anlægsbudgettering på Transportministeriets område [The main note for New construction budgeting in the area of the Ministry of Transport]. København: Transportministeriet.

Transportministeriet (2024b). Status for anlægs- og byggeprojekter Anlægsstatus 2, 2024 [Status of construction and building projects Construction status 2, 2024]. København: Transportministeriet.

Work in Denmark (2024). Workplace culture in Denmark. https://www.workindenmark.dk/working-in-denmark/workplace-culture-in-denmark (accessed 27 August 2024).

World Bank (2024). GDP per capita, PPP (current international $). World Bank.

8

Sweden

Roger Pyddoke

Swedish National Road and Transport Research Institute, Transport Economics Unit, Stockholm, Sweden

Sweden's approach to public investment governance is characterized by sector-specific systems rather than a unified national framework. In the transport infrastructure sector, the Swedish Transport Administration (TA) plays a central role, operating under a long-standing planning model that emphasizes cost–benefit analysis and goal alignment, but lacks formal stage-gate processes and systematic third-party scrutiny. Recent concerns over cost overruns and planning inefficiencies have prompted efforts to improve cost estimation and oversight. In parallel, governance of digital investments remains fragmented, with limited follow-up and coordination across agencies. This chapter explores Sweden's governance structures for transport and digital investments, highlighting key challenges, reforms, and comparisons with international practices.

8.1 Introduction: Governance of Transport Infrastructure and Digital Investments

This chapter aims to present how Swedish transport infrastructure investments are governed and what the experience with the current governance model has been. The chapter also provides some observations on the governance of digital investments in the state sector. There is no general governance system for all public investment projects in Sweden; instead, there are sectoral systems, with the system for transport investments being the one this author is most familiar with. How the current system for transport infrastructure performs can first be fully evaluated when projects planned after the current governance model was established are finished, or preferably, when they have been in use for some time.

An earlier paper that compared the governance regimes in Norway and Sweden (Olsson et al. 2019) summarized theories on the choice of bureaucratic structures (e.g. administrative procedures, rules for oversight and accountability). The governance structures analysed in this chapter are just two examples of such regimes. A main theoretical idea is that new bureaucratic structures are chosen to promote the interests of

Governance Frameworks for Major Public Projects: International Practices and Experiences, First Edition.
Edited by Vedran Zerjav, Morten Welde and Gro Holst Volden.

political parties and that these structures, therefore, also represent compromises between the parties in current majorities. An objection to this idea presented by West (1997) is that it is hard for legislators to foresee what procedures will promote their interests.

West's point is exemplified by Olsson et al. (2019) who cited evidence that projects with positive net benefit–cost ratios sometimes were not perceived as beneficial by the majority (Sager, 2016) and that emphasis on positive net benefit–cost ratios may be contrary to popularly held beliefs that new transport infrastructure is necessary to promote growth. The latter point may, therefore, place agencies in charge of promoting governance systems in conflict with ruling majorities. Holmgren (2018) indicates that this may be a considerable risk for bureaucrats, where evidence from Sweden suggests that agencies were more frequently terminated when there was conflict between enacting and sitting coalitions.

This chapter presents the Swedish governance scheme for transport investments and discusses some observations on major differences between the Norwegian and the Swedish governance of infrastructure projects. A central component in the Norwegian system is the transparent third-party scrutiny of the assessments of future costs and benefits. The advantage of these published documents is that they allow for learning about the weaknesses of assessments and the possible causes of increased costs or deficient results. The Swedish transport projects still lack full, open, third-party documented scrutiny in the early stage, the ex ante stage, and the ex post stage.

The most important criticisms of the current Swedish planning system have been that (1) cost overruns remain a problem and that (2) this allows projects that appear to give good net benefit–cost ratios to enter plans that later turn out to have low net benefit–cost ratios. This has prompted the Swedish government to encourage the Transport Administration (TA) to start a programme to improve cost calculation and control. After the initiation of the TA's programme to improve cost calculations and the commissioning of Transport analysis (TAN) (Trafikanalys, 2024) to monitor the programme, studies at the TA have resulted in confirmation that early cost estimates were further from outcome costs than late cost estimates (Trafikverket, 2024).

The contribution of this chapter is primarily an extension of the analysis in Olsson et al. (2019) up to 2017, with developments in Sweden after 2017. While Olsson et al. (2019) emphasized a comparison between Sweden and Norway, this chapter contributes a summary and synthesis of the grey literature that has examined the Swedish planning process after 2017. Finally, this chapter includes some aspects of the governance system for digital investments.

8.2 Presentation of the Model

8.2.1 Brief Introduction

The main structure of the Swedish long-term planning model has been constant since at least the 1990s. The components are as follows:

i) **Planning in two channels:** Portfolio and individual projects.

ii) High degree of delegation from ministry to agency but the degree varying over time. However, the share of projects chosen by the political level, typically with low net benefit-to-cost ratios, appears to have increased.
iii) **Two stages of portfolio planning proceeding:** First, overarching planning and, second, more detailed consideration of which projects are to be included in the plan.
iv) The planning of individual projects has been subject to several adjustments and currently consists of several steps for a successively more detailed and accurate definition of the project and precise cost calculations. These are, however, not to be compared with stage gates, as they are not formally regulated.

The legislation regulating the current planning process was described in a white paper (Prop. 2011/12:118). Four main reasons for reforming the overarching planning process in 2011 were given: integrating processes; integrating international (mainly European Union [EU]) perspectives; better cooperation between the different levels of public administration in Sweden: municipal, regional, and state, as well as cooperation with the private sector; and the need for guidelines for alternative funding. In the description of the first reason, increasing costs were emphasized (p. 61). The government also saw a continuous disruption of existing plans due to delays and cost increases and, therefore, stated a wish to be more frequently informed about developments (p. 72).

Regarding planning regulation for individual projects, some adjustments were inserted into the existing laws for road building (Väglagen; Vägförordningen, SFS 2012:707; Väglagen, SFS 1971:948) and railroad building (om byggande av järnväg; Lagen 1995:1649). The white paper (Prop. 2011/12:118), however, also gives guidance not formalized in legislation. One important new such item was the requirement to do *choice of concept studies*, which is largely equivalent to the Conceptual Appraisal in the Norwegian model. These studies, the cost estimations and the *cost–benefit analysis*, are covered by internal guidelines in the TA, but have no basis in legislation.

The need for the development of competence or education for this planning was not mentioned in the white paper. As far as I am aware, the methods used by the TA are not systematically taught at Swedish universities. However, the general methods of *cost–benefit analysis* and *models for traffic forecasts* are taught at some universities.

The overarching orientation for policy (inriktningsunderlag) is an assignment given to the TA from the government (Swedish Government, 2023). In the assignment for this work, the government states a framework for this analysis. This consists of an indicative budget (a total for the whole portfolio), projects that the government wants to be included in the portfolio, and some indication of what kind of policy instruments it wants to be included. The TA is also responsible for planning for the attainment of all relevant goals for transport policy, including climate goals (Swedish Government, 2023, p. 2).

An important part of this assignment was, therefore, goal achievement, including the attainment of climate goals. This could invite the interpretation that the government wants the TA to tentatively design a package of policy instruments (e.g. fuel taxes, blending mandates, and congestion charges) to reach the goals. This interpretation is, however, not likely since the assignment on page 6 explicitly states that the plan should be based on a scenario constructed with enacted and announced upcoming policy instruments. The TA is, therefore, not invited to design policy packages that realize Sweden's national goals.

Furthermore, the government states that welfare economic net benefit (or positive net benefit–cost ratio) should be guiding TA's prioritization of planning projects, while the government itself frequently ignores this criterion. For example, in the same assignment, the government requires that infrastructure should be provided and maintained everywhere where Swedes live. In many cases, this goal comes into conflict with the requirement of a positive net benefit–cost ratio.

The government disapproves of the fact that costs for the projects included in the previously established long-term plan 2018–2029 have experienced huge cost increases not foreseen in the plan. The government, therefore, expects the TA to prioritize and undertake extensive work to counteract cost increases wherever possible. In particular, the government wants the TA to focus on getting the cost calculations in the early planning stages closer to the outcome (Swedish Government, 2023, p. 4).

In the following stage of portfolio planning, the TA is assigned to present a detailed long-term plan of investments (Nationell plan för transportinfrastrukturen). This detailed long-term plan is for 12 years, but the plan is presented in three sections. The plan for the first phase is intended to be made up of individual projects that have already received approved building start decisions. In the basis for the detailed long-term plan, there are also *total assessments* (samlade effektbedömningar), which are documents that, in addition to the required cost–benefit analyses, also include assessments of nonmonetized consequences. In the second four-year section of the 12-year plan, projects included should also have gone far in the process of preparation for building start and preferably also have approved building start decisions. In the last four-year section of the plan (first phase in Figure 8.1), projects have been accepted that were not fully prepared, i.e. individual projects frequently have been in early stages. In the assignment for the most recent plan (Swedish Government, 2021), the government requires the TA to present a plan for all investments that the government has chosen on political grounds.

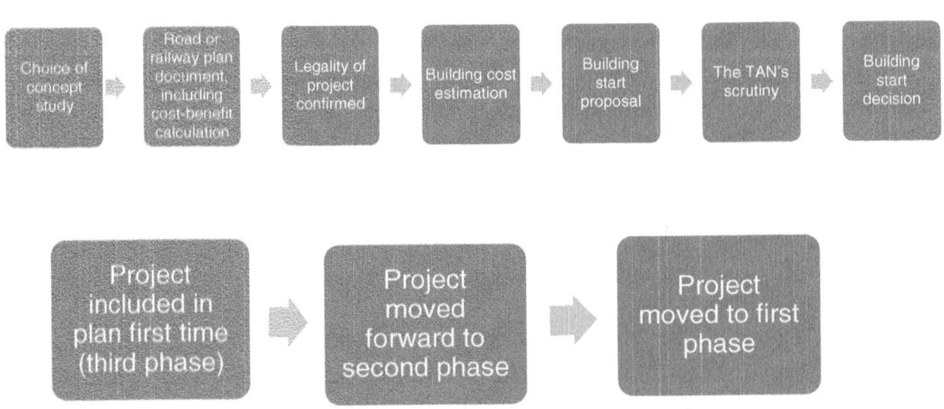

Figure 8.1 The planning process of national transport infrastructure in Sweden – Two parallel processes. Upper part: for individual projects. Lower part: for the long-term (portfolio) plan. Before a project is included in the plan for the first time, it should preferably have a building start decision. In practice, projects have been introduced solely based on the choice of concept studies.

This plan implicitly assumes that the costs will be in keeping with the most recent cost estimates. The government and the parliament steer the execution of the plan through the yearly appropriations of the budget and decide which new projects may start. If a project is delayed or costs overrun the budget, the TA will be forced to postpone the completion of some projects to keep the yearly budget. From 2023, a new building start report has been added to this process (Trafikverket, 2023a).

8.2.2 Project Initiation

My impression is that most new transport investment initiatives have come from local governments (municipal and regional) and that some of these were also directly or indirectly from private firms. When a project is perceived to be beneficial to local interests, political support may take off, and local politicians start courting national politicians and the TA.

8.2.3 Institutions and Roles

The government and the Ministry for Infrastructure's role vis-à-vis the TA are primarily to give the overarching directives in the form of the legislation regulating the planning and building and the directives for the long-term planning of roads and railroads. The Ministry, however, also identifies larger projects to be included in the plan. These projects constitute a significant share of the total budget. Furthermore, the government gives instructions and yearly assignments and budgets for the agencies. The Swedish ministries do relatively little in terms of open documented short-run monitoring of agency performance or management. This leaves more detailed allocation of budgets for smaller projects and maintenance tasks to the TA.

Some assignments are given to other agencies, for example the one given to TAN, to analyse the performance of the TA cost calculation program. As far as I know, there is also little focus on the accountability of individual civil servants, and the same seems to be the case in the other Scandinavian countries. There has recently been an initiative in the parliament to reintroduce some form of individual responsibility, but this has not been met with much approval.

There is, therefore, no 'Project management office', neither at an overarching level nor in the TA. At the TA, there is, however, a 'cost control programme' within the organization to coordinate the various sub-projects.

8.2.4 Option Appraisal

In the white paper from 2012 (Prop. 2011/12:118), the issue of early-phase planning was noticed and led to the proposal of a new kind of pre-study: the *choice of concept study*. The choice of concept study is intended to follow the so-called *four-step principle*. The four-step principle is not strictly regulated in law but rather suggested in the government's proposal (Prop. 2011/12:118) (Trafikverket, 2017, p. 10). The use of the four-step principle implies that the problem is analysed to find out if it would be possible to reduce the demand for capacity instead of expanding it. In the proposal, the emphasis was, however, mostly on the need for cooperation and co-planning with municipalities and regions, which is claimed to have led to improved cooperation between the TA, regions, and municipalities (according to Trafikverket, 2017 and Tornberg and Odhage, 2018).

The appraisal of alternatives is primarily done in the *choice of concept study* (åtgärdsvalsstudie). The guidelines for these studies have a clear focus on creating 'an arena for early dialogue in the early stages' (Trafikverket, 2015). The focus of the *choice of concept study* was more on bringing in stakeholders and producing a reasonable compromise than on producing reliable cost estimates. The do-nothing option is mentioned in the guidelines, but the guidelines are silent on how ambitious cost estimations are intended to be made and cost calculation is barely mentioned. Negotiations over funding are not included in the task of the choice of concept study.

8.2.5 Who Bears the Risk?

If a project at the TA encounters a cost overrun, this is absorbed by the TA. This means that other projects are delayed so as to reach the budget set by parliament. This leads to less room for new projects in upcoming budgets. Explicit compensation from the Ministry of Finance and the Parliament for cost overruns is unheard of. Ultimately, this leads to slower improvement of the transport infrastructure and, therefore, less realization of the benefits associated with this infrastructure.

8.2.6 Quality Assurance

The white paper also observed the need for a second opinion on the basis for building start decisions (Prop. 2011/12:118, p. 74). This was later formalized as a regular assignment to the TAN. Today, this constitutes the only part of a public infrastructure investment decision process in Sweden where a third-party regularly scrutinizes a part of the decision process. This scrutiny focuses on whether the TA complies with directives from the government and the TA's own guidelines. Such requirements include if the TA presents all relevant indicators, such as if the project fulfils all legal requirements to be started, net benefit–cost ratios, and accounts of uncertainty. This is the closest likeness to the quality assurance in Norwegian system (QA2). The Swedish scrutiny process is more rule oriented and less management oriented than the Norwegian process. Items scrutinized by the TAN include if land-use plans have gained legal approval, if cost calculations were recently updated, if calculations were based on recent forecasts and calculation values, if net benefit–cost ratios were presented, if information on contributions of alternative funding are presented, and if uncertainties are satisfactorily presented. Earlier *total assessments* (samlade effektbedömningar) were also scrutinized, but I have not been able to find recent scrutiny of these documents.

8.3 Experience with the Model

8.3.1 Culture and Traditions

Recently, the Ministry for Infrastructure expressed concerns about cost overruns in infrastructure (Swedish Government, 2023, p. 4). However, it is difficult to discern how strong the desire to curb cost overruns really is. Is it strong enough to motivate politicians to retreat from existing deals, as with earlier agreements to include projects in the plan in the face of huge projected cost increases?

Unfortunately, there are no good studies documenting the recent trends in cost overruns in Sweden and comparing these to other countries. A study by Lundberg et al. (2011), however, indicated that the average cost overruns in Sweden were modest compared to magnitudes presented for other countries. This study was, however, limited to the comparison of late-stage cost calculations with cost outcomes.

8.3.2 Transparency and Trust

The transparency of the planning system is both high and low. A substantial part of the long-term planning and the key figures characterizing projects are openly available. At the same time, the basis for the planning and outcomes of individual projects is generally not easily accessible. It does, for example, not apply to the cost estimates, which are protected by the secrecy act and have an unclear decision-making process. This means that even the TA itself has difficulties in collecting and providing data needed for a comprehensive analysis of projects.

Trust can be said to exist from the government towards the TA concerning long-term planning. The criticism from other stakeholders has mainly targeted forecasting assumptions and the choice of scenarios for demonstrating how policy instruments can be used to reach goals. For the planning of individual projects, there are no obvious indicators of large distrust. The persistent problems of delays and cost overruns have, however, repeatedly led the government to impose regulations. Some major projects have also been questioned by environmental organizations.

8.3.3 Cost Estimates in the Early-Phase (Choice of Concept) Studies

In the choice of concept studies, the net benefit–cost calculation is done in a way that undershoots the cost outcome by, on average, 33% for road projects (p. 27) and 41% for rail projects (p. 53) (Trafikverket, 2024). This does not exclude that the choice of concept analyses has other benefits in terms of reconciling the different stakeholder requirements (Tornberg and Odhage, 2018). The last stage of project design and cost estimation is the preparation for building start – at this stage, the cost estimates are closer to cost outcomes. The cost outcomes overshot cost estimates by on average 3% for road projects (p. 50) and 5% for rail projects (p. 62) (Trafikverket, 2024). The TA's study has emphasized the importance of gaining better data and understanding of cost outcomes.

8.3.4 Third-Party Scrutiny of Building Start Documents

In the third-party scrutiny of the building start documents conducted by TAN in 2023 (Trafikanalys, 2023), two important deficiencies were noted in the building start documents. The first is that the planning guidelines require that the project must have received the legal go-ahead to be included in plans, which several of the proposals do not have. The second is that the risk assessments are not sufficiently transparent. This leads TAN to question the internal quality assurance at the TA 'quality assurance of the documents and the process to produce the decisions' (Trafikanalys, 2023, p. 32).

8.3.5 Compliance with Planning Regulations

There appears to be no legal obligation to estimate costs or benefits from investments. The obligation to calculate appears to be a convention established by repeated demands from the government. Neither is there a legal obligation to provide quality assurance of estimates or planning documents.

Concerning adherence to orders and rules, the Swedish National Audit Office (SNAO) noted that assignments from the government were not strictly enforced, and the deficiencies observed by TAN do not appear to have had much effect, as many of the complaints are repeated in a later scrutiny (Trafikanalys, 2023). The SNAO (Riksrevisionen, 2023, p. 47) has also noted that the *total assessments* (samlade effektbedömningar) were frequently based on cost calculations that have been replaced by later calculations and that this is not clearly indicated. In response to criticism from the SNAO and the government regarding persistent cost overruns, the TA decided to launch a programme to improve its capability to do cost calculations. In addition to this initiative, the government commissioned TAN to follow and scrutinize the work done by the TA. In a subproject, the progress in creating supporting IT systems for cost follow-up and cost calculations at the TA has been studied (Ridderstedt and Sjöstrand, 2024). The main observations were that these development projects have been fraught with problems, but that some progress has been made.

How strictly/flexibly are these proceedings followed (Trafikverket 2017 and Riksrevisionen, 2023)? Some indications are:

i) The TA presented ex post calculations that were not openly quality checked.
ii) The SNAO (Riksrevisionen, 2023, p. 27) stated that the procedures for calculations were strict. The reviewer from the TA noted that it was correct that calculation procedures are strict for cost–benefit analysis but emphasizes that the routines for handling cost calculations were not. An important drawback is that the responsibility for cost calculations was anonymized.
iii) The SNAO (Riksrevisionen, 2023, p. 46) stated that the TA had not followed the rules for reporting large deviations to the government.
iv) Later, the SNAO (Riksrevisionen, 2023, p. 49) concluded that the government's steering of the TA's planning was vague and gave the TA room for interpretation.
v) In a report by the TA (Trafikverket, 2017), the planning process of new projects was examined. This report indicated that an intended tightening of the laws enacted in 2013 had produced both intended and unintended effects in the preparation of project plans (e.g. pages 40 and 107).

8.3.6 Ex Post Follow-Ups

Ex post follow-ups and evaluations remain a weak point in the system. Since 1997, the road and rail administrations have accounted for cost outcomes of new infrastructure projects in their annual reports. This data was later used by the Swedish Institute for Communications Analysis (SIKA, 2002) to analyse cost outcomes. The underlying documentation has, however, not been publicly available. The SNAO has examined outcomes (ex post analysis) several times (Riksrevisionsverket 1994, Riksrevisionen 2010, 2011). In 2010, the SNAO (Riksrevisionen, 2010) examined road investments. The findings were that overruns had continued. The reported cost increases in the annual reports systematically underestimated the true cost increases. Furthermore, the cost outcomes were only compared to the latest cost calculations (the building start calculations). The SNAO found no systematic follow-up of cost deviations and their causes (Riksrevisionen, 2010, p. 9). Risk assessments were also criticized. The TA's principal method for risk assessment in cost calculations – successive calculation – was deemed unfit for purpose (Riksrevisionen, 2010, p. 10). Again, in 2011 the SNAO (Riksrevisionen, 2011) reported on railroad investments. The observations were essentially parallel to those presented earlier for roads.

The most recent analysis by the SNAO of cost control in the TA (Riksrevisionen, 2021) presented several criticisms based on the theme of lacking transparency of changes in content and costs, as well as a lack of analysis of cost outcomes and their causes. Implicitly, this reveals that the SNAO holds the normative stance that the resources for such analyses should be increased at the TA. Surprisingly, the SNAO referred to its own reports from 2010 and 2011 but refrained from commenting on the parallel between the 2010 and 2011 critiques, on the one hand, and the most recent critique of 2021, on the other hand, and the lack of improvement. The SNAO instead emphasized that the TA had established a database for saving ex post calculations within a project and a 'routine for saving calculations' (p. 29). This routine has, however, not been applied retroactively. The SNAO also noted that the TA had difficulties in collecting calculations for ongoing projects. There is also still no database or routine for comparison of projects.

Therefore, large cost increases have prevailed despite the attention given to the issue by SNAO. The SNAO has pointed out that transparency is reduced due to the difficulty in collecting previous cost estimates and outcomes. The loss of transparency is not, however, total since the SNAO has collected data on a few occasions to make comparisons possible.

8.4 Some Observations on the Governance of Digital Investments

The Agency for Digital Government (Myndigheten för digital förvaltning, DIGG) was formed in 2018. The agency was given the mission to 'coordinate and support common digitalization investments for public administrations to make public administration more efficient and fit for purpose' (SFS 2018:1486). Exceptions were investments made by the ministries and the defence sector.

In 2020, DIGG and the Swedish Agency for Public Management (APM, Statskontoret) were commissioned by the government to propose new forms of governance for digital investments in public administration. A first report (Statskontoret 2020) reviewed the status of the internal governance of state agencies' investments in digital infrastructure and the government's governance of the agencies. The agency directed its investigation to agencies with long-term experience with digitalization and some with shorter-term experience (21 agencies in total). The APM found that most agencies had made investments in digital infrastructure and that they had worked systematically. The prime motive for these investments was improved service quality, but some also aimed for lower costs.

The most important weaknesses lay in the follow-ups. The documentation of the conditions before the investment and the priority for follow-up were weak. A frequent reason for not having pursued follow-ups was that these require resources and that the benefits are accrued outside the agency.

Concerning the government's governance of the agencies, the APM concluded that government oversight and steering gave no clear view of how the government wants its agencies to work with digitalization. The government lacks an 'overview of digital investments' and that the creation of DIGG was intended to provide an overview; however, DIGG has not been able to provide this (Statskontoret 2020, p. 6). The APM also noted that the 'government had posed few requirements on agencies to report benefits and effects of digitalization related assignments or internal digitalization' (p. 6). When such requirements were posed, these were directed at large agencies. In particular, for the governance of joint investments involving more than one agency, there was a need for strengthened governance from the government (p. 7).

In the final report (DIGG 2020), the following proposals for government governance were given:

- For investments involving more than one agency, the government should provide goals and promote collaboration.
- The government should also establish a judicial unit in the ministry and provide a budget for the joint funding of investments involving more than one agency.
- The government should, to a larger extent, require agencies to account for benefits from investments, and DIGG guidelines should be updated to include more of an economic perspective.

A prime reason for the Norwegian project governance system requiring external quality assurance is that it provides incentives for doing good *cost–benefit analysis* and for improving by learning about weaknesses in the assessment of future effects and costs. Compared to Norway, two aspects stand out in the Swedish context. First, the recommendations from DIGG do not reveal an awareness of the fact that cost overruns or poor net benefit–cost ratios may be an issue. Second, there is no clear demand for a third-party perspective on investments or external assessment of investments.

8.5 Conclusion

Delays and large cost overruns persist in Swedish transport infrastructure investments, despite the scrutiny of the SNAO. The implementation of project governance lags in Sweden, both in transport infrastructure and in digital investment. The important part lacking in Sweden is the open third-party scrutiny of plans and cost–benefit analyses. A lack of historical perspective on the efforts to reduce cost uncertainty and overrun risks results in the loss of useful experience.

The lag in implementation is unlikely to be caused solely by a lack of knowledge that project governance systems of the Norwegian type exist. One hypothesis is that it is the experience of uncomfortable information from cost–benefit analysis that reduces the political enthusiasm for stricter project governance. This ties back to theories of democratic institutional design, where institutions emerge that are perceived to promote stakeholder interests. The issue of introducing firmer discipline coming with a more formal project governance system ultimately rests on governing parties' perceptions of the net benefits of such discipline.

Acknowledgements

I thank Gro Holst Volden, Peo Nordlöf, Mattias Lundberg, and two anonymous reviewers for their comments. Funding was provided by NTNU and VTI.

References

DIGG (The Swedish Agency for Digital Government) (2020). *Styrning av digitala investeringar, Dnr: 2020–325*. Sweden: Stockholm.

Holmgren, M. (2018). Partisan politics and institutional choice in public bureaucracies: evidence from Sweden. *Journal of Public Administration Research and Theory* 28. https://doi.org/10.1093/jopart/muy010: 355–370.

Lundberg, M., Jenpanitsub, A. and Pyddoke, R., 2011, Cost overruns in Swedish transport projects, CTS Working Paper 2011:11

Olsson, N.O.E., Nyström, J., and Pyddoke, R. (2019). Governance regimes for large transport infrastructure investment projects: comparative analysis of Norway and Sweden. *Case Studies on Transport Policy* 7: 837–848.

Prop. 2011/12:118. *Planeringssystem för transportinfrastruktur (Planning system for transport infrastructure)*, white paper to the Swedish parliament.

Ridderstedt, I. and Sjöstrand H. (2024). *Koll på kostnaderna – Kartläggning av Trafikverkets digitala infrastruktur för kostnadskontroll*, VTI rapport 1219.

Riksrevisionen (2010). *Kostnadskontroll i stora väginvesteringar*? Rapport 2010:25.

Riksrevisionen (2011). *Kostnadskontroll i stora järnvägsinvesteringar*? Rapport 2011:6.

Riksrevisionen (2021). *Kostnadskontroll i infrastrukturinvesteringar*, Rapport 2021:22. Summary in English available at the website www.riksrevisionen.se.

Riksrevisionen (2023). *Nationell plan för transportinfrastrukturen– lovar mer än den kan hålla*, Rapport 2023:25. Summary in English available at the website www.riksrevisionen.se.

Riksrevisionsverket (1994). *Infrastrukturinvesteringar en kostnadsjämförelse mellan plan och utfall i 15 större projekt inom Vägverket och Banverket*, Rapport 1994:23.

Sager, T.Ø. (2016). Why don't cost-benefit results count for more? The case of Norwegian road investment priorities. *Urban Planning and Transport Research* 4 (1): 101–121.

SIKA (2002). *Metoder och riktlinjer för att förbättra samhällsekonomiskt beslutsunderlag*, Rapport 2002:19.

Statskontoret, (The Swedish Agency for Public Management) (2020). Styrning av digitala investeringar, Dnr 2020/40-5, Stockholm, Sweden.

Swedish Government (2021). *Ministry of Infrastructure, Uppdrag att ta fram förslag till nationell plan för transportinfrastrukturen och möjlighet att ta fram länsplaner för regional transportinfrastruktur*, Dnr: I2021/01904.

Swedish Government (2023). Ministry of Rural Affairs and Infrastructure, Uppdrag att ta fram inriktningsunderlag inför den långsiktiga infrastrukturplaneringen för planperioden 2026–2037, Dnr: LI2023/02737.

Tornberg P. and Odhage, J. (2018). Meningen med gemensamma planeringssammanhang – En studie i nyttan med åtgärdsvalsstudier, Institutionen för Samhällsplanering och miljö, KTH.

Trafikanalys (2023). *Granskning av Trafikverkets byggstartsförslag 2023*, Rapport 2023:6.

Trafikanalys (2024). *Granskning och uppföljning av Trafikverkets arbete med kostnadskontroll – delredovisning*, Rapport: 2024:5.

Trafikverket (2015). *Åtgärdsvalsstudie – nytt steg i planering av transportlösningar*.

Trafikverket (2017). *Utvärdering av planläggningsprocessen – Erfarenheter av 2013 års Lagstiftningsförändringar*, by Witzell, J., Publikationsnummer: 2017:091.

Trafikverket (2023a). *Byggstartsrapportering 2023 Kompletterande redovisning*.

Trafikverket (2024). *Utveckla* Referensklassprognoser: Delrapport Trafikverkets referensklasser år 2024.

West, W.F. (1997). Searching for a theory of bureaucratic structure. *Journal of Public Administration Research and Theory* 4: 591–613.

9

The United States

George Edward Gibson

School of Sustainable Engineering and the Built Environment, Arizona State University, Tempe, AZ, USA

The governance of public investment projects in the United States is shaped by its federal structure, with distinct roles for federal, state, and local entities. While no single unified model exists, federal agencies such as the General Services Administration (GSA), Federal Highway Administration (FHWA), Department of Energy (DOE), and U.S. Army Corps of Engineers (USACE) follow structured stage-gate processes rooted in front-end planning principles. These frameworks are designed to ensure transparency, accountability, and performance across a vast and diverse portfolio of infrastructure and capital projects. This chapter explores how these agencies implement governance practices, highlighting the role of federal regulations, budgeting processes, and quality assurance mechanisms in shaping project outcomes.

9.1 Context and Policy Basis

The United States is a large and economically dominant country with a population exceeding 335 million. It remains the world's largest economy by nominal GDP and ranks among the top 10 globally in GDP per capita (PPP), which stood at approximately US$85,810 in 2024.[1]

The U.S. economy is highly diversified, driven by advanced sectors such as technology, finance, healthcare, and energy. Public investment remains substantial, particularly in infrastructure, clean energy, and digital transformation, supported by recent federal initiatives such as the Infrastructure Investment and Jobs Act and the CHIPS and Science Act.

The federal system comprises three tiers: federal, state, and local governments. While the federal government collects the majority of tax revenue, states and municipalities play a critical role in delivering public services and infrastructure. This decentralized structure means that a complex interplay of federal incentives and local priorities often shapes public investment decisions.

1 https://data.worldbank.org/indicator/NY.GDP.PCAP.PP.CD?locations=US.

Governance Frameworks for Major Public Projects: International Practices and Experiences, First Edition.
Edited by Vedran Zerjav, Morten Welde and Gro Holst Volden.

The American work culture emphasizes individual accountability, performance, and innovation. While this fosters dynamism and entrepreneurship, it can also lead to fragmented decision-making and limited coordination across agencies. Transparency is supported by robust freedom of information laws, though access to public data varies significantly across jurisdictions.

This chapter mainly focuses on the construction of public sector buildings and infrastructure to discuss the governance scheme. In the U.S. context, the governance scheme will be discussed in relation to practices of 'programme and project management', 'capital budgeting and management processes', or 'acquisition structure and strategy' (Federal Highway Administration (FHWA) 2024a; General Services Administration (GSA) 2024b; U.S. Department of Energy (DOE) 2024). Governance of projects is typically carried out through a stage-gate project delivery process.

When viewed from abroad, the United States is sometimes considered to be homogeneous in its approach to public project management. This is not a realistic view, as public expenditures are overseen differently at the national (federal) level, state level, and local municipality level and widely dispersed across a broad continental landscape and in some cases internationally.

Much of the Federal public documentation and governance requirements used today were established over a century ago, evolved over time, and have been widely used and modified as the country moved into support of national infrastructure and weapons programmes for the Second World War and later. For further reading, see Coats and Brady (2015).

This chapter focuses on the U.S. Federal-level model, which only applies to projects where federal monies are spent (highways, buildings, major procurements, waterways, laboratories, and military projects). There is a great variety in implementing project governance at different levels and segments of the public sector and for different types of projects. As a result, a single project model could be identified, but a myriad of models are used and implemented by a variety of organizations that deal with government projects.

The chapter is mainly focused on the insights and perspective of the Construction Industry Institute (CII) long-standing research stream on the front-end planning (FEP) processes in the development of major projects (Construction Industry Institute (CII) 2013a,b, 2017). Major projects in construction represent a significant part of government capital expenditures spread across the federal level, state level, county, and municipalities.[2] The author has had significant involvement in this stream of research. Similarly, the author has considerable familiarity with project approvals and governance schemes for agencies and departments at the federal level, which plan, design, and construct facilities for the U.S. Federal Government.[3] Those that were selected as illustrative examples for governance scheme practices are the *General Services Agency (GSA), (FHWA), DOE, and the U.S. Army Corps of Engineers (USACE).* Each of those is responsible for a portfolio capturing a significant portion of the annual U.S. Government capital facilities investments and is representative of Federal executive agencies and departments.

---

2 US public expenditures on construction projects amount to about $483 billion USD annually, representing 23% of the construction expenditures across both public and private sector (US Census Bureau 2024).

3 Examples include the U.S. Architect of the Capitol, Smithsonian Institution, Naval Facilities Command, U.S. Department of Health and Human Services, and many others.

Table 9.1 Overview of the reviewed agencies.

	Agency			
Information	**General Services Administration (GSA)**	**Federal Highway Administration (FHWA)**	**Department of Energy (DOE)**	**U.S. Army Corps of Engineers (USACE)**
Major types of projects	Construction and operations of U.S. governmental buildings	Maintenance and construction of highways	Large science, energy, nuclear projects, and environmental clean-up	Dams, waterways, infrastructure construction, and operations
Presidential budget (2023)	US$10.6 billion	US$73 billion	US$33 billion	US$27.3 billion

Insights presented here are based on the CII extensive research on the FEP processes and practices from over 400 organizations and completed projects worth more than US$125 billion since 1991 (Construction Industry Institute (CII) 2013a,b, 2017). A brief overview of these agencies is provided in Table 9.1. These expenditures represent roughly 30% of U.S. public capital expenditures during 2023.

The GSA is tasked with building and maintaining federal buildings, providing workspaces for all other government agencies. It has several branches such as Federal Acquisition Services (FAS) and Public Building Services (PBS), which work together in order to provide for the office building needs of the government (General Services Administration (GSA) 2024a).

The FHWA is an independent division within the U.S. Department of Transportation that manages and builds the interstate roads across the country. The operations of FHWA are then divided into two main subcategories of the Federal-aid Highway Program and the Federal Lands Highway Program. The Federal-aid Highway Program is in place to give financial allowances and funds in support of state and territory projects that involve building roads and bridges (Federal Highway Administration (FHWA) 2024a).

DOE oversees maintaining and addressing the energy and environmental needs of the United States. DOE responsibilities are research-heavy, and because of that, the largest system of national laboratories belongs to this department. DOE has three offices that address most of the largest projects, nuclear security, science and energy, and environmental management, with other programme offices and staff functions. Projects are constructed both through the Laboratory system and weapons production sites, mostly in the United States, and overseen by the respective sponsors with quality assurance (QA) by the Office of Project Management (OPM) (U.S. Department of Energy (DOE) 2024).

Established over 200 years ago, USACE is the engineering arm of the U.S. Army and has the responsibility of maintaining and constructing infrastructure, primarily locks, dams, and canals in the United States, military construction projects worldwide (through Department of Defence funding) and managing the nation's waterways from an environmental perspective. It also performs combat engineering missions for the regular army (U.S. Army Corps of Engineers (USACE) 2024).

To ensure the quality and transparency of projects, all governmental procurements of GSA, along with FHWA, DOE, and USACE, are regulated by the Federal Acquisition Regulations (FAR), which are under chapter 1 of 48 Code of Federal Regulations (CFR) (General Services Administration (GSA) 2024a). The CFR is a document that contains all the regulations in place for the executive departments of the United States. It contains principal rules set for all agencies under 50 main titles with several chapters for each to direct public departments in their duties. For example, Title 10 of CFR covers Energy, regulating the operations of the DOE in two chapters (e-CFR 2020).

Many agencies exist under the executive branch of government at the federal level and can expend capital underneath an overarching U.S. model and set of regulations for acquisitions (FAR) and budgeting (Office of Management and Budget (OMB) Circular A-11), but each has latitude to modify their model within these frameworks based on their unique mission (U.S. Government 2024b; Office of Management and Budget (OMB) 2024). Each agency in the Federal model has its approval stage-gate model, but the projects are then further approved by the OMB (executive branch) and Congress (legislative branch). Hence, the process at the U.S. Federal level contains at least three QA steps.

GSA, along with the Department of Defence and NASA, issues the FAR for the rest of the public agencies, except a few exempted departments, to use as a guideline for procurement of goods and services (U.S. Government 2024b). The FAR system was developed to codify uniform policies and procedures for acquisition by all executive agencies. The FAR system consists of the FAR, which is the primary document, along agency's acquisition regulations that implement or supplement the FAR.

9.2 Overview of the Project Governance Schemes in Selected Agencies

Given that the U.S. country context differs from the other country settings of this book, the structure of presenting the governance scheme is slightly different. To this end, we first outline the capital budgeting and allocation process in each of the federal agencies considered before detailing the stage-gate process with its milestones.

9.2.1 Budget Allocation Process

9.2.1.1 General Services Agency

GSA issues contracting opportunities for the development of projects proposed by their clients to several GSA-registered contractors. After the initial review of the offers and per the requirements set by the FAR, the award goes to the lowest responsive, responsible bidder in accordance with the regulations. Prospectus development studies (PDS) are reviewed from submittals of each region, and after examination of their deliverability and requirements, GSA sends requests to the Congress for funding. It is only after the approvals of OMB and meeting cost estimates standards per GSA benchmark that a project can begin its execution phase. Environmental assessment and acquisitions take place in the beginning of the execution process, and constructability reviews are submitted to GSA HQ. GSA runs several front-end analyses on each project to ensure the best project is selected. Throughout the development phase, several project reviews are done to ensure adherence to the budget and progress plans.

9.2.1.2 Federal Highway Administration

FHWA requires each state to set aside a percentage of their budget to allocate sufficient funds to each transportation programme per fiscal year (Federal Highway Administration (FHWA) 2024b). Local departments of transportation must then match federal funds with 10% of the expenditures. When a project is estimated at more than US$500 million, FHWA rates it as a major project and takes an active leadership role. About 25% of U.S. roads qualify as major projects. Projects under US$500 million are managed by local entities with oversight by FHWA. All acquisitions made by FHWA are regulated by Transportation Acquisition Requirements (TAR), a supplementary entity to Title 48 of the CFR and FAR (U.S. Department of Transportation (U.S. DOT) 2024). It should be noted that there are stage-gate reviews at several points in this process, most notably after Project Concepts are developed, again after Preliminary Design is completed and at the end of Final Design, with each having a cost estimate review at those points. For major projects, FHWA usually requires an initial cost estimate review of all alternatives within 90 days of proposal. The cost estimation review workshops continue throughout the project development phases to support the financial plan (FP). Initially, it is submitted to the appropriate division (usually the local department of transportation) for approval before receiving federal funding. FHWA projects usually have more than four checkpoints throughout the development (U.S. Department of Transportation (U.S. DOT) 2024).

9.2.1.3 Department of Energy

The DOE oversees maintaining and addressing the energy and environmental needs of the United States. Project management requirements and capital asset acquisition rules are set by DOE HQ and the OPM at DOE, per Order 413.3B with the stage-gate process. DOE Order 413.3B also includes guidelines on QA, risk management, FEP, execution planning, cost estimations and performance baselines (PBs) for projects greater than US$50 million (U.S. Department of Energy (DOE) 2024). Critical Decisions (CDs) correspond to stage-gate approvals in these processes. Projects above this US$50 million threshold are required to report performance to the Project Assessment and Reporting System (PARS), establish an Earned Value Management System, and are overseen by the OPM. Projects are subject to external independent reviews (EIR) at various points along the process. The Energy System Acquisition Advisory Board (ESAAB) has been set up to assist the Department, and meetings are scheduled to aid the senior acquisition executive in planning a decision-making process and in making CDs.

9.2.1.4 U.S. Army Corps of Engineers

USACE has divisions across its operational territories within the continental United States and other parts of the world. These Divisions are further broken into Districts, which oversee and manage planning, design, construction, and, in some cases, operations. Besides, these divisions are sub-organizations that enhance the operations of the USACE. For instance, the U.S. Army Engineering and Support Centre (CEHNC) oversees construction and project management. CEHNC bids on several federal construction

Table 9.2 Comparison of the schemes at four agencies.

	Criteria			
Agency	**GSA**	**FHWA**	**DOE**	**USACE**
Portfolio management	GSA + the U.S. Congress	FHWA + Local Jurisdic-tions + the U.S. Congress	DOE + the U.S. Congress	USACE + DOD + the U.S. Congress
Who decides the choice of concept?	Federal agencies + GSA	FHWA + local jurisdictions	DOE + sub-organizations[a)]	USACE + Department of Defence (DOD) + local jurisdictions (sometimes)
Determination of the budget	GSA region	FHWA districts	Laboratories, as well as NNSA, OS, EM	USACE districts and divisions
Threshold value[b)]	US\$1–US\$5 million	US\$500,000	US\$20 million	US\$2–US\$26 million depending on type
Project appraisal[c)]	GSA agents	Local DOT	Office of Science + PM	Sponsor/district appraiser
Who performs quality assurance	CQM through the agency itself	CQM through the agency itself	CQM through the agency itself	CQM through the agency itself
Project budget calculation	Internal + external	Internal + external	Internal + external	Internal + external
Transparency	Inspector General + GAO	Inspector General + GAO	Inspector General + GAO	Inspector General + GAO
Acquisitions	GSA	FHWA + local jurisdiction	DOE	USACE + DOD
Typical co-funding option/requirements	None	10% at the state level	None	None

CQM: Construction Quality Management is the performance of tasks, which ensure that construction is performed per plans and specifications, on time, within a defined budget, and safe work environment; DOD: Department of Defense; DOT: Department of Transportation; GAO: General Accountability Office.

a) Department of Science (OS), National Nuclear Security Administration (NNSA), Environmental Management (EM), etc.
b) No reviews necessary if the value does not exceed the prospectus threshold (41 CFR §102.703.35).
c) All federal appraisals are done per UASFLA (Uniform Appraisal Standards for Federal Land Acquisition) guidelines.

programmes by GSA and other federal agencies and plans those projects in accord with USACE Engineering Regulation (ER) 25-1-8. Most of the USACE projects are planned as Design-Build (DB) or Design-Bid-Build (DBB). Design reviews are performed at 65%, 95%, and 100%, requirements and goals are set based on the USACE engineering manuals and protocols. Per ER 1105-2-100, a feasibility report is required by USACE, and District commanders are the entities responsible for approving the technical quality of proposals. QA reports are mandatory parts of the USACE project development plan and are reported to the districts for review of policy compliance (U.S. Army Corps of Engineers (USACE) 2000, 2024). Table 9.2 provides an overview and comparison of the four agencies described earlier. Each follows a similar set of processes, but the mission requirements of each are very different. Funding thresholds are set by the agencies and vary by project type or focus. Projects below those thresholds are still subject to the FAR and are also expected to be managed efficiently, even though funding decisions are made at the local level and are typically subject to total budget thresholds given to the local management team.

U.S. Federal agencies decide on whether a project progresses forward to OMB and Congress, as well as overseeing the project stages for QA, and once approved by OMB and Congress, the Office of President does not have direct authority over the outcome of a particular project. In general, U.S. Federal agencies operate with delegated liberties in terms of decision-making and QA. However, budget guidance from the Executive Branch and local officials (Congress) will have significant influence on which projects are approved and funded.

FEP and adherence to a stage-gate process develop the foundation for predictable and efficient project delivery, helping organizations to meet their project drivers while accomplishing mission priorities. The National Research Council (NRC) highlighted the need for effective FEP of the DOE projects (National Research Council (NRC) of the National Academies 2004). Research further indicates that well-performed FEP reduces costs and variances and increases the likelihood of meeting project goals. Additionally, DOE conducted a root cause analysis (RCA), indicating the importance of FEP on project management performance (U.S. Department of Energy (DOE) 2011).

It is also important to mention that above the agency level is the executive and congressional oversight, which serves as an additional QA layer. For brevity purposes, however, this layer is not covered in this chapter.

9.2.2 Overview of the Stage-Gate Project Models

Stage-gate models followed by the agencies described earlier are shown in Figure 9.1. Each project and agency is subject to external assessments (conducted by entities outside the agency, such as the U.S. General Accountability Office, which works for Congress) and internal assessments (conducted by an Inspector General, an independent entity tied to the Agency). Additionally, each agency has internal QA functions that hire consultants to perform independent cost estimates. Each generally also employs independent external and internal reviews of progress as the project progresses through its life. Key funding decision points are subject to technical reviews of project efficacy and, of course, political assessment in many cases.

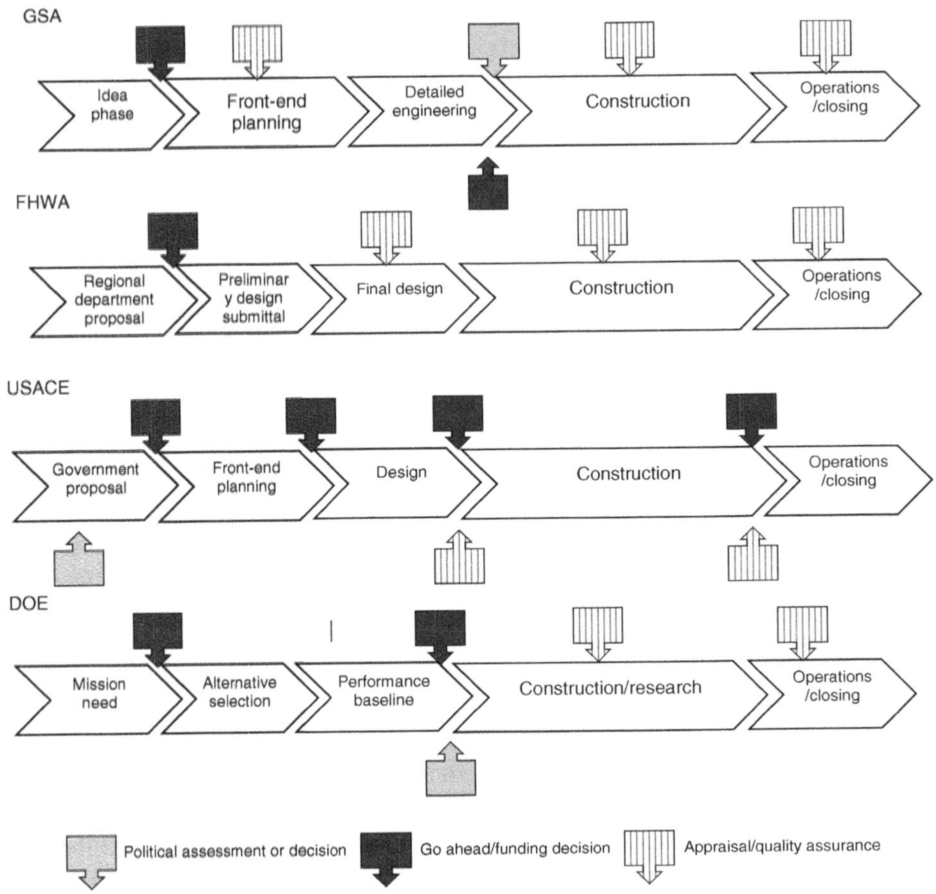

Figure 9.1 Summary of the reviewed U.S. Agencies' stage-gate model.

Stage-gated FEP processes are widely used on both public and private projects in the U.S. effective FEP practices and a stage-gate process followed for project planning, and approval process are synonymous as the process allows for effective involvement of the key stakeholders and the parties to systematically understand and approve CDs during the early life cycle of the project (Construction Industry Institute (CII) 1997; Gibson and Pappas 2003; Gibson et al. 2022). The level of decision-maker at each stage-gate in the process is typically related to the size of the expenditure or the level of risk of the project. Effective evaluation and decision-making are critical at these junctures of the project.

CII's research results showed performance for both public and private projects in the United States and abroad has been similar, with those projects adhering to a structured and disciplined process, overall, being more successful. The Project Definition Rating Index (PDRI) has been adopted and used by many public agencies over the years, including GSA, NASA, DOE, U.S. Department of Health and Human Services, Smithsonian Institution, U.S. Architect of the Capitol, and USACE. In addition, much of its terminology has been adopted into standard practices across the U.S. Government.

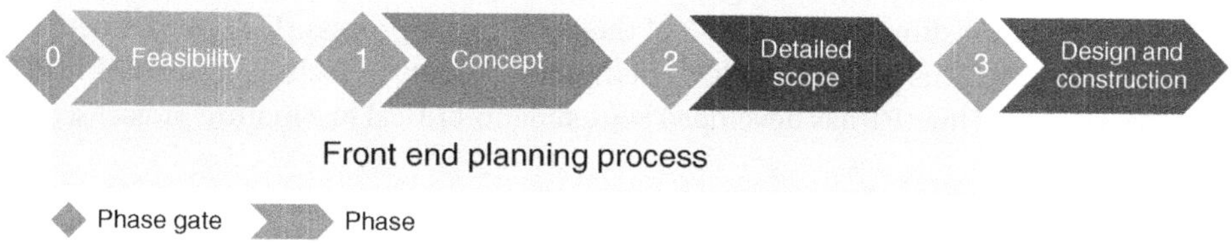

Figure 9.2 High-level FEP stage-gate process with phases and gates. *Source:* Construction Industry Institute (CII) (2019) / CII.

Less than effective FEP leads to less than desired results. *Maturity* in this context refers to the completeness of engineering deliverables at the end of the FEP phase (Gate 3 in Figure 9.2). It assesses whether the necessary documents, designs, and specifications are sufficiently developed to support detailed design and execution. *Accuracy*, on the other hand, measures the confidence in the maturity assessment – how reliable and trustworthy the deliverables are, based on the processes and personnel involved in their creation. A recent CII study showed that projects with a highly mature and accurate (environment in which the project is being established) front-end engineering design (FEED) outperformed projects with low maturity and accuracy by 24% on cost performance (Construction Industry Institute (CII) 2019).

CII's PDRI tools developed an effective method for measuring the degree of scope development for construction projects across a wide spectrum of projects, helping integrated project teams (IPTs) as they move through the stage-gate process. The six PDRI tools offer comprehensive lists of 64–70 scope definition elements (depending on the large project tool) divided into three key areas for project planning: Basis of Project Decision, Basis of Design, and Execution Approach. Each element is grouped within categories by topological similarities and is weighted based on its relative importance to the other elements. A scoring scheme through the project stages of development allows the users to evaluate the state of completeness of scope definition at any point prior to detailed design and construction; and where the scoring identifies gaps, it allows the project team to quickly predict factors impacting project risk. Since the PDRI score is related to risk, those areas (elements within the categories, such as Site Information) that need further work can easily be identified. CII empirical studies have shown that an overall score of 200 (300 or below for the small project tools) or less prior to determining the project baseline can greatly increase the probability of a successful project (U.S. Department of Energy (DOE) 2023).

FEP and adherence to a stage-gate process develop the foundation for predictable and efficient project delivery, helping organizations to meet their project drivers while accomplishing mission priorities. The NRC highlighted the need for effective FEP for DOE projects (National Research Council (NRC) of the National Academies 2004). Research further indicates that well-performed FEP reduces costs and variances and increases the likelihood of meeting project goals. Additionally, DOE conducted an RCA, indicating the importance of FEP on project management performance (U.S. Department of Energy (DOE) 2011).

In 2023, DOE published a guide that provides their federal project directors (FPDs) and IPTs with the information, methodologies, and tools to perform effective front-end project planning before establishing a PB at CD-2 of their stage-gate process (U.S. Department of Energy (DOE) 2023). This guide includes a discussion of tools to include in the CII PDRI and to assess whether the IPT has developed key elements critical to a mature project scope definition.

The DOE guide includes the CII PDRI for Industrial Projects version 5.0, which incorporates the PDRI Maturity and Accuracy Total Rating System (PDRI MATRS) and other DOE-developed tools based on the PDRI. CII research over the years has found that FEP is a socio-technical system. The accuracy of any plan is statistically related to the 'environment' in which the team is working. The CII work initially developed a team alignment measurement and then later a more rigorous way to measure this environment through a rating system (Griffith and Gibson 2001; Construction Industry Institute (CII) 1997).

9.3 Experiences Using the Governance Scheme

As indicated earlier, different agencies and different government levels operate autonomously, and in the absence of a unitary governance scheme, it is difficult to discuss outcomes of the governance schemes used in project and programme development. However, CII has undertaken several studies using a wide sample of U.S. construction organisations to understand how the quality of FEED impacts projects. This system emphasizes two critical dimensions – maturity and accuracy – as essential indicators of project readiness and predictors of success. The findings of this stream suggest that those projects with both a mature technical approach to FEP and a good team environment performed much better statistically (24% better in cost performance versus final estimate) than projects that did not (Yussef et al. 2018, 2019). The accuracy rating method assesses factors related to the owner project team, the contractor project team, the project management process and practices being followed, and the resources allocated to perform the FEP process itself (Construction Industry Institute (CII) 2017).

This author feels that overall, compliance is done well, but it is uneven as the total programme is so large and distributed, and decision-making is sometimes heavily influenced by politics. From the outside, this process appears bureaucratic and not agile, but it generally safeguards against unwise investments and systematic fraud. It does have many checks and balances, which are justified.

Most U.S. Federal agencies employ effective FEP processes revolving around stage-gate decision points. Many are moving toward integrated project and programme management tied to a project control and management system with a strong foundation built around an effective project scope.

It should be noted that an effective stage-gate process is a socio-technical system consisting of the proper level of design, risk management, cost estimating, scheduling, and constructability along with a team that has the right amount of expertise at the owner and contractor level, as well as adequate resources and good practices. Marrying effective FEP to integrated project/programme management results in delivering projects on time, on budget, and meeting sponsor requirements.

Acknowledgements

The author would like to acknowledge the effort of Pedram Esmailzadeh, who began initial work on this topic as a Master's student in 2020 and developed some of the diagrams included in this chapter.

References

Coats, D. and Brady K. (2015). An economic history of federal spending and debt. https://www.jec.senate.gov/public/_cache/files/aeeff50d-dc8e-4e0b-ab9f-def32f184179/20150910-jec-spendingstudy.pdf (accessed 10 September 2015).

Construction Industry Institute (CII), Gibson, G.E. (coordinating author) (1997). Alignment during pre-project planning; a key to project success. The Construction Industry Institute, Implementation Resource 113-3, December, 84 pp.

Construction Industry Institute (CII), Gibson, G.E. and Bosfield, R. (2013a). Assessment of effective front end planning processes. The Construction Industry Institute, Research Summary 268-1a, 38 pp., October.

Construction Industry Institute (CII), Gibson, G.E. (coordinating author), Podesta, T., Fish, J., Halicks, D., Bosfield, R., MacGillivray, S., Carlsson, E., Werle, B., Cooley, D., Foy, B. Yuengert, J., Mitrocsak, R., Bryson, K., and Scott, R. (2013b). Adding value through front end planning. The Construction Industry Institute, Special Publication 268-3, 36 pp., January.

Construction Industry Institute (CII), El Asmar, M. and Gibson, G.E. (coordinating authors) (2017). Front end engineering design (FEED) maturity and accuracy total rating system (MATRS). The Construction Industry Institute, Implementation Resource 331-2, 142 pp., October.

Construction Industry Institute (CII), Gibson, G.E., El Asmar, M., and Cho, A. (2019). Project definition rating index industrial projects, A Front End Planning Maturity and Accuracy Total Rating System (PDRI MATRS), The Construction Industry Institute, Implementation Resource 113-2, Version 5.0, 224 pp., July.

e-CFR (2020). Electronic code of federal regulations. https://gov.ecfr.io/cgi-bin/ECFR?page=browse (accessed 4 April 2020).

Federal Highway Administration (FHWA) (2024a). Major projects. https://www.fhwa.dot.gov/majorprojects/ (accessed 18 October 2024).

Federal Highway Administration (FHWA) (2024b). Project development and design manual. https://highways.dot.gov/federal-lands/pddm (accessed 26 October 2024).

General Services Administration (GSA) (2024a). Acquisition policy. https://www.gsa.gov/policy-regulations/policy/acquisition-policy?topnav=policy-regulations (accessed 23 October 2024).

General Services Administration (GSA) (2024b). Project planning guide. https://www.wbdg.org/FFC/GSA/projectplanningguide.pdf (accessed 31 October 2024).

Gibson, G.E. and Pappas, M. (2003). Starting smart: key practices for developing scopes of work for facility projects. Federal Facilities Council Technical Report #146, 87 pp., December. National Academies Press. https://doi.org/10.17226/10870. ISBN: 978-0-309-16,668-3.

Gibson, G., El Asmar, M., Sanboskani, H., and Aramali, V. (2022). Implementing the integrated project/program management (IP2M) maturity and environment total risk rating (METRR) using EVMS in a team environment. Report 6, School of Sustainable Engineering and the Built Environment, Ira A. Fulton Schools of Engineering, Arizona State University, 192 pp., July. https://doi.org/10.13140/RG.2.2.20392.85764.

Griffith, A.F. and Gibson, G.E. (2001). Alignment during pre-project planning. *ASCE Journal of Management in Engineering* 17 (2): 69–76.

National Research Council (NRC) of the National Academies (2004). *Progress in Improving Project Management at the Department of Energy, 2003 Assessment*. Washington D.C.: The National Academies Press.

Office of Management and Budget (OMB) (2024). Circular No. A-11, preparation, submission and execution of the budget. https://www.whitehouse.gov/wp-content/uploads/2018/06/a11.pdf (accessed 15 September 2024).

U.S. Army Corps of Engineers (USACE) (2000). Planning guidance notebook. https://www.publications.usace.army.mil/Portals/76/Publications/EngineerRegulations/ER_1105-2-100.pdf (accessed 20 April 2000).

U.S. Army Corps of Engineers (USACE) (2024). Planning community toolbox. https://planning.erdc.dren.mil/toolbox/index.cfm (accessed 3 November 2024).

U.S. Department of Energy (DOE) (2011). Root cause analysis and corrective action plant closure report. https://www.energy.gov/projectmanagement/articles/root-cause-analysis-and-corrective-action-plan-closure-report (accessed February 2011).

U.S. Department of Energy (DOE) (2023). Front-end planning and project definition rating index for nuclear and non-nuclear construction projects. DOE-G-413.3-12-Chg. 2 (27 September 2023).

U.S. Department of Energy (DOE) (2024). DOE Order 413.3B Chg. 7. https://www.directives.doe.gov/directives-documents/400-series/0413.3-BOrder-B-chg7-ltdchg (accessed 25 October 2024).

U.S. Department of Transportation (U.S. DOT) (2024). Transportation acquisition regulations. https://www.acquisition.gov/tar (accessed 18 September 2024).

U.S. Government (2024). Federal acquisition regulations (FAR). https://www.acquisition.gov/far/part-1#FAR_1_000 (accessed 15 October 2024).

US Census Bureau (2024). Monthly construction spending. https://www.census.gov/construction/c30/pdf/release.pdf (released 2 September 2024).

Yussef, A., Gibson, G., El Asmar, M., and Ramsey, D. (2018). Front end engineering design (FEED) for large industrial projects: FEED maturity and its impact on project cost and schedule performance. Construction Research Congress 2018, American Society of Civil Engineers, New Orleans, LA (May 2018).

Yussef, A., Gibson, G.E., El Asmar, M., and Ramsey, D. (2019). Quantifying FEED maturity and its impact on project performance in large industrial projects. *ASCE Journal of Management in Engineering* 35 (5). https://doi.org/10.1061/(ASCE)ME.1943-5479.0000702: 2019.

10

Canada (Federal Government)

Olivier Choinière[1], Maude Brunet[2] and Nathalie Drouin[3]

[1]*Department of Management Sciences, Université du Québec à Rimouski, Lévis, QC, Canada*
[2]*Department of Management, HEC Montréal, Montréal, QC, Canada*
[3]*Department of Management, Université du Québec à Montréal (UQAM), Montréal, QC, Canada*

The federal Canadian government has a long-standing history of project governance. The first policy on major public investment projects was introduced in 1978 with the publication of the *Policy and Guidelines on the Management of Major Crown Projects* (Treasury Board of Canada 1978). Since its inception, this policy framework has constantly evolved, with each version addressing new issues ranging from governance structure and administrative rules to decision-making processes and its scope of application, leading to the implementation of increasingly prescriptive instruments involving a diverse array of stakeholders (Brunet and Choinière 2024a).

10.1 Context and Policy Basis

Canada is the second-largest country in the world, by land mass, covering nearly 10 million square kilometres with a population of slightly over 40 million people, the majority of whom are concentrated near the United States border. This vast territory, combined with a growing yet ageing population, is exerting pressure on most of the public infrastructure across the country (Organisation for Economic Co-operation and Development 2024a).

Canada's economy has experienced a strong recovery from the pandemic downturn, with a relatively low unemployment rate alongside workforce shortages in key sectors such as healthcare, construction, and high technology (Organisation for Economic Co-operation and Development 2024b). This recovery has led to high inflation, prompting the Bank of Canada to swiftly increase its overnight interest rate from near zero to 5% over a 16-month period before gradually reducing it (Bank of Canada 2024). As a result, the cost of major projects mirrored these interest rate fluctuations. Other factors, such as global supply shortages, rising input costs, and other international agreement challenges, have also affected the steady increase in project costs.

The Canadian governmental landscape consists of three main levels of government: the federal jurisdiction based in Ottawa, 10 provinces and three territories, and thousands of

Governance Frameworks for Major Public Projects: International Practices and Experiences, First Edition.
Edited by Vedran Zerjav, Morten Welde and Gro Holst Volden.

municipalities. The Constitution of 1867 (and its amended version of 1982) outlines the executive powers granted to each level of government, as well as those shared between them, particularly in areas such as transportation and public infrastructure; however, the separation of constitutional responsibilities between Canadian governments often remains open to interpretation, leading to multiple debates and tensions among the various political entities (Champagne and Choinière 2016). This constitutional ambiguity is a central element in the definition, negotiation, approval, and financing of major public projects, which usually fall under shared jurisdictions (Siemiatycki 2015).

Concretely, the federal government either assumes ownership of major public projects through the framework described in this chapter or funds megaprojects that fall under provincial and municipal jurisdictions, guided by constitutional competencies and negotiation dynamics. These projects follow a specific policy governing transfer payments selection and approval. Funding is granted and monitored through various federal entities, notably the Canada Infrastructure Bank and the Department of Infrastructure Canada (Government of Canada 2021a).

10.2 Presentation of the Framework

The federal governance framework comprises a series of integrated institutional arrangements across various policy levels, including a law, a policy, a directive, and two standards (Government of Canada 2008). At the first level, the *Financial Administration Act* provides the legal foundation for the sound management of administrative functions, public funds, and performance standards. This Act also defines the role and authority of the Treasury Board (TB) (Government of Canada 1985).

At the second level, the *Policy on the Planning and Management of Investments* outlines the roles and responsibilities of deputy ministers, along with the expected outcomes regarding the assets and capabilities each department requires to deliver products and services to Canadians effectively. Key management functions addressed by this instrument include programme and project management, real property, information technology, materials, finance, human resources, security, and legal services (Government of Canada 2021a).

The third level features the *Directive on the Management of Projects and Programmes,* which provides formal instructions to the deputy heads and their subordinates. This instrument details effective governance and controls throughout the programme and project life cycles, emphasizing decision-making at each stage gate. It also states the importance of gathering performance measurement data for ongoing monitoring and evidence-based decision-making. Furthermore, the Directive mandates that departments develop and maintain their capacity in line with organizational needs (Government of Canada 2019).

The fourth level consists of standards that provide mandatory and detailed procedures and practices, such as the Organizational Project Management Capacity Assessment (OPMCA) and the Project Complexity and Risk Assessment (PCRA), which are discussed later in this chapter. Finally, a selection of guidelines and tools is available to offer voluntary advice and guidance to project managers and practitioners.

Collectively, these instruments emphasize governance and control, aiming to optimize efficiency and ensure the realization of project benefits. This suite of policy instruments articulates the government's expectations for monitoring projects and programme health, evidence-based decision-making, and evaluating project and programme management capacity, tailored to meet departmental needs and maturity levels (Government of Canada 2019, 2021a).

In summary, the federal governance framework establishes the foundation for selecting major projects under federal ownership. It also requires each department to develop its own framework, as detailed in the 'Stage-Gate Model' section.

10.2.1 A Brief Historical Overview

This specific suite of federal policy instruments was developed over a span of four decades. The initial iteration of the framework was published in 1978, following several high-profile major crown projects – referred to as major projects under federal responsibility – that were considered less than successful, such as the Mirabel airport (located in the Greater Montreal Area) and the purchase of F5 fighter jets. At the time, departments managed projects 'in a non-coordinated manner, with costs sometimes exceeding their annual operating budget' (Treasury Board of Canada 1978).

Over the following decades, successive Liberal and Conservative governments updated the framework according to their political preferences, yet remained constrained by institutional limitations.

Following the initial iteration (1978–1985), which focused solely on major crown projects, the second version (1985–1994) introduced two more instruments: the *Management and Control of Projects* and the *Project Management and Training*, along with an update on the 1978 policy. During this period, the federal government began to question the existing economic model of previous administrations, embracing the privatization of crown corporations such as Air Canada and Petro-Canada. This exit strategy, as well as a decentralization of federal powers, significantly reduced the portfolios of projects owned by the government.

The third iteration (1994–2009) resulted in the creation of two new policies: the *Project Management Policy* and the *Project Approval Policy*. The former instrument was largely driven by the urgent need to fight against rising public debt, as illustrated by its official objective to 'achieve effective and economical management of projects with visible and clearly established project leadership' (Government of Canada 1994). The latter instrument focused more on the approval controls and mechanisms for specific project areas, such as capital lease and information technology (Government of Canada 2005).

The fourth iteration (2009–2019) consolidated and enriched previous instruments into the *Policy on Management of Projects*, notably adding two key standards detailed: the *Standard of Organizational Project Management Capacity* and the *Standard for Project Complexity and Risk*. The focus of this revamped policy was value for money, resource optimization, and accountability.

The current iteration, the fifth since its inception, has been in effect since 2019. It continues the trajectory toward heightened project management responsibilities and requirements. The project governance aspect has been integrated into a broader policy framework that includes seven other policies, while the management aspect has been scaled down to a directive, focusing more on 'how' rather than 'what' (Brunet and Choinière 2024a).

10.2.2 The Stage-Gate Model with Phases and Decision Points

As part of the project's governance policy suite, the Treasury Board Secretariat (TBS) introduced in 2021 an updated stage-gate model to assess project viability and continued investment worthiness (Government of Canada 2021b, p. 2). It is important to note that the stage-gate process is conducted by the government and usually does not require any promulgation by the Parliament.

This generic model outlines the following six key gates through a project's life cycle, providing a 'formal review of a project's health' to inform the decision regarding its continuation (*idem*):

- **0**: *Verify business problem or opportunity (pre-project)*: This initial phase focuses on defining the problem, articulating the desired state, and validating the concept case. This gate is numbered zero as the project has not yet been formalized.
- **1**: *Validate business justification and select short list of options*: In this second step, a comprehensive list of options is developed and subsequently narrowed down to a preferred option that demonstrates the desired benefits, justifying its selection along with a preliminary cost analysis. The successful passage of this gate signifies that the project is officially initiated.
- **2**: *Approve preferred option and approach*: At this step, the business case is developed, along with a more comprehensive cost–benefit analysis, the project approach, and delivery strategy. The project then receives funding to refine the project definition further.
- **3**: *Fund project deliverables*: This step aims to obtain expenditure authorities for implementation (based on the cost–benefit analysis); a project management plan is developed, procurement activities are carried out, and the projected benefits are agreed upon.
- **4**: *Confirm readiness for service*: Following the development of the project deliverables, this gate ensures that the project has provided the necessary capacity and capability for transitioning the product or service to operations, including testing and training.
- **5**: *Complete project and conduct post-implementation review*: To successfully pass this gate, all project activities must be finalized, and any outstanding issues, along with accountability for benefits realization, are assigned to operations.

The guide stipulates that there must be 'one individual accountable for making gating decisions'; however, consultation is required. This individual must gather evidence to support their decision. For example, the project sponsor typically consults with the chair of the investment board (these roles are explained in the 'Institutions and Roles' section) (Government of Canada 2021b, p. 12).

Based on the model provided by TBS, departments need to adapt the central model using four layers of instruments. At the organizational level, the *Departmental Project*

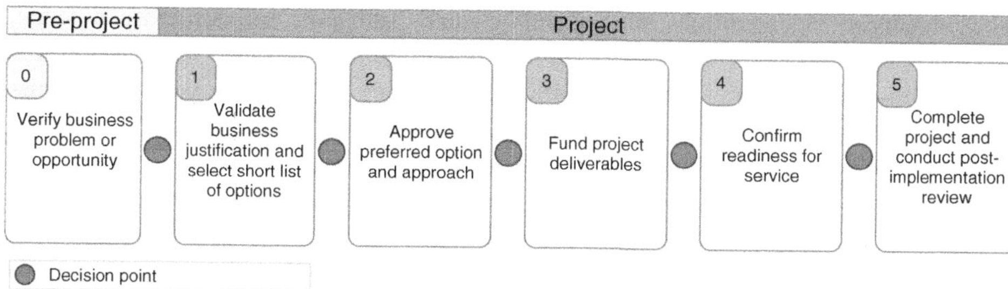

Figure 10.1 The federal stage-gate model in Canada.

Management Framework outlines the governance process, systems, and controls. The *Departmental Project Gating Model*, overseen by the Senior Designated Officer, specifies the key project decision points and requirements at each gate to approve the release of pre-approved funds. Each model needs to identify the appropriate number of gates, as there is no TBS requirement for the total number of gates, and outline the criteria that must be fulfilled to advance through each gate as 'departments are encouraged to tailor their gating models to their unique circumstances' (Government of Canada 2021b, p. 9). Moreover, a department may have more than one gating model depending on the size, risk, and complexity of its projects (Government of Canada 2021b, p. 6).

At the project level, the *Project Management Plan* details key elements such as scope, schedule, cost, risks, issues, outcomes, and benefits, while the *Project Gating Plan* applies the gating model specifically to a project under the supervision of the project sponsor. These plans are established during the Treasury Board Submission or before business case approval, with updates possible throughout execution. The same gating approach applies to hybrid and Agile projects, which may progress through specific gates several times (Government of Canada 2021b, p. 8).

Each gate results in one of four possible decisions: continue investment without conditions, impose new conditions before advancing to the next gate, request resolution of outstanding issues before moving forward, or simply terminating the project (Figure 10.1).

10.2.3 Projects That Must Adhere to the Model

In theory, all federal projects must adhere to the model based on specific expenditure thresholds detailed later. These projects vary widely in nature, from infrastructure initiatives like the new Gordie Howe International Bridge (connecting Windsor, Canada to Detroit, USA) to military procurement like the National Shipbuilding Strategy, a portfolio of projects aimed at renewing the fleets of the Canadian Royal Navy and Coast Guard, as well as real-estate projects such as the Parliament Hill renovation initiative, the largest and most complex heritage building renovation in Canadian history (Government of Canada 2023a).

Every department identified under section 7 of the *Financial Administration Act* must implement its own project governance framework based on the government model. This includes over 100 departments, agencies, and crown corporations that cover unique or shared federal responsibilities, such as public works, transport, infrastructure, defence, and Indigenous services. However, agents of Parliament (e.g. Office of the Auditor General, Office of the Chief Electoral Officer, and Office of the Information Commissioner of Canada) are not subject to this model, as they operate independently from the government's executive power.

Departments that adhere to the government project governance framework must conduct two mandatory assessments: *OPMCA* and *Project Complexity and Risk Assessment (PCRA)*. The expenditure authority of each project is determined by the aggregate results of these appraisals. Currently, there are four thresholds of expenditure authorities: CA$2.5 million, CA$5 million, CA$10 million, and CA$25 million. For projects exceeding these thresholds, departments must request the appropriate authorization from the TB, which entails a higher degree of scrutiny from TBS. Additionally, departments must submit the following reports to the Office of the Comptroller General for every project over CA$25 million: project approval, expenditure authority, amendments, and project close-out (Government of Canada 2019).

In detail, each department must conduct an OPMCA on a three-year cycle and submit the results of its analysis to TBS, which then determines the organization's classification according to a five-level maturity model (limited, sustaining, tactical, evolutionary, and transformational). The evaluation is based on a self-administered questionnaire (92 questions) covering three core capacities and 13 areas of project management knowledge (listed in Table 10.1). Supporting documentation must be provided to substantiate the assessment. This process is typically carried out by the Departmental Project Management Office (DPMO), which is responsible for gathering evidence from project leads and assigning a score to each question. The evaluation process, which can take several months to complete, requires sustained commitment from stakeholders

Table 10.1 Organizational Project Management Capacity Assessment instrument.

Core capacity	Areas of project management knowledge	Number of questions
Organizational integration	Investment portfolio management	10
	Investment program management	
	Organizational support structures	9
	Project management standards	20
Core project management	Integration management	21
	Scope management	4
	Time management	5
	Cost management	7
	Risk management	4
Supporting project management	Quality management	2
	Procurement management	4
	Human resource management	4
	Communications management	2

Source: Adapted from Government of Canada (2013).

Table 10.2 Project Complexity and Risk Assessment instrument.

Categories	Number of questions	Sample questions
Project characteristics	18	What percentage of the total project cost estimate is for procurement?
Strategic management risks	6	How well and how clearly does the project align with the organization's mandate and strategic outcomes?
Procurement risks	9	What is the supplier availability and willingness?
Human resource risks	5	Does the organization anticipate a shortage of available personnel with appropriate skills during a significant period of the project?
Business risks	5	What is the level of public involvement required to achieve expected outcomes?
Project management integration risks	6	Have feasibility studies been conducted, and is there confidence in the assumptions made in the feasibility studies?
Project requirements risks	15	To what extent are the project's requirements clear, completed, and communicated?

Source: Adapted from Government of Canada (2015).

due to the volume of consultation and evidence gathered, as well as the interactions between the department and TBS, which oversees the compliance and integrity of the exercise.

In conjunction with the evaluation of organizational capacity, assessing each project's complexity and risks enables the establishment of authorized spending limits and the level of oversight from central agencies (minimal or significant). The assessment of complexity (based on project size, administrative rules, and technology used) and risks should ideally be conducted during the project definition phase and can be updated as the project evolves and more information becomes available. The tool comprises 64 questions divided into seven categories, as presented in Table 10.2.

The results of this exercise classify the project into one of four levels: sustained, tactical, evolutionary, and transformational. Therefore, the more complex and riskier a project, the higher the organizational capacity in project management required to avoid sustained oversight from TBS. Conversely, if a department implements projects that are generally low risk or low complexity, the organization do not need to invest heavily in enhancing its project management capacities.

Beyond these assessments, the federal government aims to provide departments with tools to make informed decisions and investments, thereby enhancing their capacity to manage projects effectively and efficiently in an integrated manner. However, these exercises require significant time and resources.

10.2.4 Institutions and Roles

Several actors play key roles in developing and implementing the governance framework, both at the central and departmental levels.

During the political appraisal phase, major projects are typically included in the federal budget through the budget lines of the sponsoring departments, which is tabled in the House of Commons. Once the budget receives royal assent (the equivalent of a promulgation in a parliamentary monarchy), departments must submit a Treasury Board Submission – highlighting the content of stage-gates 0–3 – to TB to obtain project approvals and spending authorities.

After securing an initial, high-level political approval, the technical appraisals are conducted between the sponsoring department and central agencies, such as the TBS and the Comptroller General. These appraisals are guided by expenditure authorities, which are determined by both governmental and departmental governance arrangements. The following section provides a brief overview of these actors and their primary roles within the framework.

10.2.4.1 Governmental Project Management Framework

The stakeholders at the governmental level include a committee of selected ministers, central agencies, and selected departments based on their mandate. While the involvement of each actor may vary depending on the project's triple constraints, certain institutions, such as the TB and its Secretariat, are engaged in every major project.

- The **Treasury Board of Canada** (TB) is a committee responsible for managing government operations, consisting of five ministers, including the President. TB evaluates selected projects based on submissions brought forward by sponsoring ministers to seek approval, which includes specific authorities (Government of Canada 2022).
- The **Treasury Board of Canada Secretariat** (TBS) is the administrative arm of TB. TBS plays a vital role within the framework by establishing and documenting government-wide processes; overseeing capacity, complexity, and risk assessments; and reviewing all documents sent to TB, for example the Treasury Board Submission (Government of Canada 2021a, 2024a).
- The **Office of the Comptroller General of Canada** (OCG) is tasked with providing overarching direction in financial management across the government. The framework stipulates that the OCG must offer leadership and oversight for all projects valued at over CA$25 million (*idem*).
- The **Office of the Chief Information Officer of Canada** (OCIO) offers strategic direction and leadership on information technology, management, and security, along with support and guidance on capacity building and project management for digitally enabled projects, as well as managing the digital investment oversight program (*idem*).
- **Public Service and Procurement Canada** (PSPC) is the department responsible for governmental-wide procurement activities, among other responsibilities. For acquisitions over specific thresholds, projects must conduct their acquisition process in tandem with PSPC.
- Finally, **other departments and agencies** might be integrated into the project structure according to their specific mandates, for example environmental protection, indigenous business development, global affairs, and industrial capabilities.

10.2.4.2 Departmental Project Management Framework

At the departmental level, key actors consist of either individuals or committees composed of executives. Their roles are outlined in the main policy instruments, such as the 2019 Directive on Management of Projects and Programmes and the 2021 Guide to Project Gating.

- The **Senior Designated Official** (SDO) serves as the department's project authority and is responsible for the development, implementation, and monitoring of the departmental project management framework.
- The **Project or Programme Sponsor** plays both a leadership and a governance role by being responsible and accountable for the planning, definition, implementation, transition, and close-out of the project/programme.
- The **Project Manager** is responsible for the day-to-day management of a specific project. This role involves coordinating with the Project Sponsor to align management and governance levels effectively.
- The **Departmental Investment Board** ensures governance and strategic alignment across project portfolios. Among its various responsibilities, it recommends front-end gate passages for major projects.
- The **Departmental Project Management Office** (DPMO) is typically responsible for stage gate preparation and provides advice to the Departmental Investment Board. Each department has its own DPMO, which varies in maturity and responsibilities, ranging from stage-gate preparation to offering methodological guidance to specific projects.

Figure 10.2 depicts the relationship among these governmental and departmental actors.

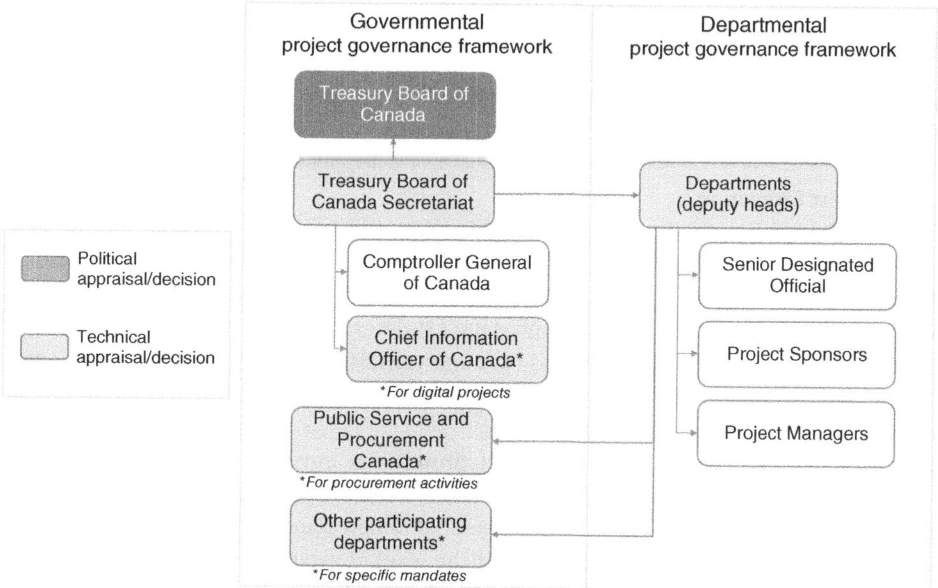

Figure 10.2 Institutions and roles involved in the federal governance framework.

10.2.5 Selected Topics

The following section focuses on selected topics that are commonly found in various government frameworks and schemes.

10.2.5.1 Front-End

The first three gates can be viewed as the front-end stages, occurring prior to the funding for the execution of the project being granted. These gates expose the business problem or opportunity (Gate 0), to validate the business justification and select a short list of options (Gate 1), and to approve the preferred option and approach (Gate 2).

The outputs of the front-end phases are a comprehensive business case, a high-level project management plan, an initial procurement plan as well as a first version of a PCRA. These, in conjunction with the OPMCA, provide the spending authorities and level of scrutiny by central agencies.

10.2.5.2 Option Appraisal

The options appraisal is central to Gates 1 and 2. As a complementary instrument, the TBS's *Business Case Guide* outlines the following requirements for this key assessment: a preliminary options analysis, key evaluation criteria, a list of possible options, screening of those options, the rationale for both discounted and viable options, and a comprehensive analysis of viable options. This examination includes considerations such as strategic alignment, costs, cost–benefit analysis, implementation, departmental capacities, risks, benchmarks, policy and standards to follow, as well as advantages and disadvantages.

The Status Quo must be included in the *Possible Options* step to illustrate how the department would perform if it did not engage in the proposed initiative (also known as the base case). This appraisal concludes with the identification of a preferred option, which will require approval before proceeding to the planning phase, i.e. stage-gate 3 (Government of Canada 2009).

10.2.5.3 Cost Estimation

Cost estimation is a crucial component of the stage gate model. The recently published Guide to Costing (Government of Canada 2023b) outlines a seven-step process: (1) establish a purpose and a plan; (2) employ a cost breakdown structure; (3) identify ground rules and assumptions; (4) gather evidence and data; (5) determine uncertainty, risk, and cost; (6) assure and validate the exercise; and (7) communicate the results. The process is not strictly linear, as some steps can be iterative when necessary: for example, steps 3–6 may need to be reassessed based on new findings.

The Business Case Guide states that cost projections must include the total cost of ownership, which encompasses ongoing expenses throughout the investment's life cycle, along with any potential compliance costs for stakeholders (Government of Canada 2009). The cost estimation is then evaluated against the level of expenditure authority based on organizational capacity, complexity, and risk assessments. Regarding contingencies, if departments need to request additional funding, they must submit a new Treasury Board Submission to TB through TBS.

10.2.5.4 Quality Assurance

This process is designed to confirm the viability, strategic alignment, and achievability of proposed outcomes and benefits of projects, thereby facilitating effective decision-making and oversight.

The governmental framework offers three types of quality assurance: quality management (daily controls), governance arrangements, and independent assurance. The first two types can be conducted by project sponsors and their teams, while an independent review is performed by qualified individuals at arm's length from project authorities, such as the internal audit function or third-party reviewers.

Independent reviews come in three forms: full review (conducted at the gate passage), quick assessment (health check), and targeted reviews (in-depth evaluation of specific critical issues). It is important to note that independent reviews are mandatory for government-wide IT projects (Government of Canada 2021a, p. 14).

10.2.5.5 Ex Post Evaluation

In the context of retrospective assessment, the Project Brief, an essential instrument part of the governance framework, mandates an evaluation strategy and a provision for an independent assessment if necessary (Government of Canada 2019). However, the requirement to conduct such an evaluation has been omitted since the conclusion of Iteration 3 (1994–2005).

Currently, Gate 5 specifies the need for a post-implementation review to validate that all project activities have been completed, marking the project's closure. It also notes that after the project's closure, the benefits of implementing the project's outputs and capabilities will continue to be realized within 'the operational (business-as-usual) environment' (Government of Canada 2021b). Consequently, the responsibility for ex post evaluation, particularly regarding the value generated by the project, falls to the internal client in charge of operations. As a result, there is a lack of publicly available evidence for these evaluations, except for reports issued by Agents of Parliament pertaining to specific projects, often those that were unsuccessful, such as the Phoenix pay system (Office of the Auditor General of Canada 2018).

10.2.5.6 Competencies

The 2019 Directive stipulates that federal project managers must possess the necessary competencies and experience for effective management (Government of Canada 2019). To support this, the government provides a range of courses on governance and project management through the Canada School of Public Service, its dedicated learning centre. One such course, the *Project Management in a Government Context*, emphasizes federal policies, directives, and procedures, as well as the 'roles and responsibilities involved in managing projects within a federal government context' (Government of Canada 2024b). Additionally, various private sector and not-for-profit organizations also offer project management courses, which lead to professional certifications such as PMI's Project Management Professional (PMP) and Prince2.

10.3 Experience with the Model

Providing real-life experience with the framework can become challenging, as appraisals are not published and must be requested through a lengthy access-to-information process, which can take months, if not years, to yield results. Additionally, documents are often heavily redacted, hindering an in-depth analysis of the model's influence over projects (Brunet and Choinière 2024b). However, agents of Parliament, such as the Auditor General of Canada, produce audits and evaluations that are accessible online and can provide valuable insights.

Nonetheless, sufficient information was available to highlight a positive experience of governing and managing a major project according to the best practices of the model, the Samuel de Champlain Bridge Corridor Project in Montreal, Québec.

The structure of the original Champlain Bridge, inaugurated in 1962, deteriorated rapidly due to construction choices coupled with harsh winters and salting. In 2015, an agreement was signed between the government of Canada and Signature on the St Lawrence (SSL) group, to cover the design, construction, financing, operation, maintenance, and rehabilitation of a new bridge. The new 3.4-km-long bridge, spanning the St Lawrence River, replaces the previous one and is designed to last 125 years. This major project was designed and constructed in 48 months, despite the severe winter conditions in the province, a six-month delay compared to the original planning. The Samuel De Champlain Bridge was officially opened on 1 July 2019 at a cost of approximately CA$4.5 billion.

This major federal project demonstrated compliance with key governance characteristics of the model, including a rational basis for decision-making, such as the delivery method, enhanced control over the private partner, and improved front-end development. Success factors of this project include: (1) a transparent procurement process; (2) the selection of the private partner; (3) an accelerated delivery approach; and (4) a clear focus on sustainable development and urban integration.

10.3.1 Remaining Challenges

As demonstrated in this chapter, the government of Canada has a long history of developing and refining its project governance framework for major public investment projects. This suite of instruments is well-established within departments and mandatory for projects that meet specific organizational maturity and project expenditure criteria, or otherwise strongly recommended.

Despite these time-tested practices, significant challenges remain. For example, a research project conducted by TBS and the University of Ottawa reveals that although departments claim to follow the framework, they often fail to deliver the expected project benefits and meet stakeholders and citizen needs (Ika et al. 2023). Several key observations emerged from the research: in many cases, authority assignment remains unclear or is not applied, while responsibilities are broadly delegated but not enforced. Over-optimism in the initial scope and lack of planning during the early stage-gates weakened the project sponsor's ability to address risks and issues, leading to delays, cost escalations, and rescoping. Furthermore, in some instances, business cases were not treated as an essential instrument to support rational decision-making, but instead as a checklist item to support funding requests. Competency gaps across all disciplines were also noted, hindering the proper implementation of the framework.

Over the past four decades, progress has been made and considerable effort has been invested in driving improvements; however, much work remains to ensure that senior leaders and project managers are effectively selected, supported, and positioned for success.

References

Bank of Canada (2024). *Statistics: Interest Rates*. Bank of Canada.

Brunet, M. and Choinière, O. (2024a). Governance framework trajectory for major public projects: the case of Canada. *Project Management Journal* 56 (1): 74–89.

Brunet, M. and Choinière, O. (2024b). Grands projets publics fédéraux: une bureaucratie de plus en plus complexe et opaque. *La Conversation* (4 September 2024).

Champagne, E. and Choinière, O. (2016). Le financement des infrastructures municipales et les défis du fédéralisme fiscal canadien. *Gestion et Management Public* 4 (3): 25–36.

Government of Canada (1985). *Financial Administration Act*. Department of Justice Canada.

Government of Canada (1994). *Project Management Policy*. Treasury Board of Canada Secretariat.

Government of Canada (2005). *Project Approval Policy*. Treasury Board of Canada Secretariat.

Government of Canada (2008). *Foundation Framework for Treasury Board Policies*. Treasury Board of Canada Secretariat.

Government of Canada (2009). *Business Case Guide*. Treasury Board of Canada Secretariat.

Government of Canada (2013). *Guide to Using the Organizational Project Management Capacity Assessment Tool (Version 1.3)*.

Government of Canada (2015). *Project Complexity and Risk Assessment Tool (Version 1.4)*. Treasury Board of Canada Secretariat.

Government of Canada (2019). *Directive on the Management of Projects and Programmes*. Treasury Board of Canada Secretariat.

Government of Canada (2021a). *Policy on the Planning and Management of Investments*. Treasury Board of Canada Secretariat.

Government of Canada (2021b). *Guide to Project Gating*. Treasury Board of Canada Secretariat.

Government of Canada (2022). *Treasury Board Submissions*. Treasury Board of Canada Secretariat.

Government of Canada (2023a). *The Centre Block Project*. Government of Canada.

Government of Canada (2023b). *Guide to Costing*. Government of Canada.

Government of Canada (2024a). *Treasury Board of Canada Secretariat Organization*. Government of Canada.

Government of Canada (2024b). *Project Management in a Government Context (TRN323)*. Canada School of Public Service.

Ika, L., Drouin, N., Grant, G., and Delcorde, C. (2023). Towards moving from outputs to outcomes delivery in major digital programs. Research reports of the major projects observatory of the Telfer School of Management, University of Ottawa (May 2023).

Office of the Auditor General of Canada (2018). *2018 Spring Reports of the Auditor General of Canada to the Parliament of Canada: Report 1—Building and Implementing the Phoenix Pay System*. Office of the Auditor General of Canada.

Organisation for Economic Co-operation and Development (2024a). *OECD Member: Canada*. Organisation for Economic Co-operation and Development.

Organisation for Economic Co-operation and Development (2024b). *OECD Indicators*.

Siemiatycki, M. (2015). Public-Private Partnerships in Canada: Reflections on twenty years of practice. *Canadian Public Administration* 58 (3): 343–362.

Treasury Board of Canada (1978). *Policy and Guidelines on the Management of Major Crown Projects*. Treasury Board of Canada Secretariat.

11

Quebec

Maude Brunet[1], Nathalie Drouin[2] and Olivier Choinière[3]

[1]*Department of Management, HEC Montréal, Montréal, QC, Canada*
[2]*Department of Management, Université du Québec à Montréal (UQAM), Montréal, QC, Canada*
[3]*Department of Management Sciences, Université du Québec à Rimouski (UQAR), Lévis, QC, Canada*

In Quebec (Canada), the first initiative to implement a policy to frame the governance of major public projects goes back to 2008 (Brunet 2019). Since then, the framework has undergone three additional iterations: first in 2010, then in 2014, and most recently in 2025 when the current version was published, the *Directive sur la gestion des projets majeurs d'infrastructure publique*. While there have been some benefits related to its implementation (Brunet 2021), its legitimacy has been regularly questioned, as major projects' costs still escalate, and numerous challenges remain.

11.1 Context and Policy Basis

Quebec is one of the 13 provinces and territories of Canada. It is the largest province by area and the second largest by population (8.9 million), as well as the northernmost province in Canada. Quebec's official language is French, and the only Francophone-majority province of Canada. Quebec is a vibrant, multicultural province with plenty of natural resources, vast forests, rolling hills, and countless waterways. The GDP per capita was $58,100 CAD in Canada in 2024 (Statistics Canada) and $48,133 CAD in Québec in 2021 (Institut de la statistique du Québec). Quebec's economy is highly industrialized and diversified.

Like other Canadian provinces, over the past 15 years or so, Québec has reinvested massively in infrastructure (Michaud et al. 2023). Upward pressure on infrastructure spending will continue, both to absorb the maintenance deficit and to build new infrastructure. For example, with a total of 31,091 km of provincial roads, the average asset maintenance deficit in dollars per capita is $2257 CAD (Michaud et al. 2023).

11.1.1 The 'Directive'

The Quebec governance framework addresses specific issues regarding regulation, decision-making processes, and its scope. After a brief historical overview of its

Governance Frameworks for Major Public Projects: International Practices and Experiences, First Edition.
Edited by Vedran Zerjav, Morten Welde and Gro Holst Volden.

developments, the main features of the governance framework are described, notably the stage-gate process, the role of actors, and the quality assurance of the projects' documentation. Now largely adopted in major public projects, as it is mandatory for projects exceeding $50 million (CAD), it has nonetheless been contested in several instances, being perceived as rigid and not well suited for collaborative delivery. A new version of the framework was issued in 2025. The new version of the framework (Secrétariat du Conseil du trésor 2025) offers greater flexibility in terms of infrastructure maintenance projects and alternative contracting methods. The goal is to reduce average construction costs by 15% and accelerate the construction of schools, roads, hospitals, and public transportation networks by 25%. The revised framework now offers greater flexibility and several possible paths, depending on (1) the nature of the infrastructure (maintenance project or improvement project); (2) the chosen delivery method; and (3) whether the project is subject to environmental impact assessment and review procedure.

11.1.2 A Brief Historical Overview

The origin of the Quebec governance framework can be traced back to a report from the Auditor General of Québec about the poor performance of the project of the extension of the Montreal subway to Laval (Vérificateur général du Québec 2004). The main recommendation of this report was that the government adopt a governance framework for major public projects, based on a rigorous planning process and on the best practices from the field. Following this report, the Treasury Board Secretariat (*Secrétariat du Conseil du trésor*, SCT) of Quebec undertook important work with the help of the Center for Interuniversity Research and Analysis of Organizations (CIRANO). The CIRANO team was led by researcher Roger Miller, who was responsible for the IMEC research programme (International programme in the Management of Engineering and Construction), an international partnership conducted from 1995 to 1998 for which 60 projects served as an in-depth comparative study at an international scale to identify best practices in project management (Miller and Lessard 2000). The CIRANO team produced a series of reports, 11 in total, in order to be used by SCT to produce a governance framework for major public projects. Those reports included many recommendations regarding the framework, structures, and project portfolio, among other things (Miller et al. 2006a,b). One of the reports was a benchmarking of governance frameworks, in which the Norwegian and British frameworks were predominant and served as examples to follow (Castonguay and Miller 2006). The SCT issued in 2008 its first governance framework for public projects: *La Politique-cadre des grands projets d'infrastructure publique* (Secrétariat du Conseil du trésor 2008).

At the time, public–private partnerships (PPPs) were widely used to conduct major projects in Quebec. The Quebec Agency for PPPs, created in 2005, had an important role to play in the governance framework for public projects at the time. However, due to major public resistance, persistent criticism of this mode of realization for major projects, and the controversial projects of the construction of mega hospitals in Montreal, the government announced in 2009 the end of this organization, to be converted to Infrastructure Quebec in 2010. The Politique-cadre was accordingly revised in 2010, in order to downgrade the predominant role of PPP projects to one of the possible modes of realization for projects (Secrétariat du Conseil du trésor 2010). The role of Infrastructure Quebec was defined at that time and made central to the coordination of the governance framework for projects. However, despite the adoption of this framework and its revision in 2010, a 2011 report by the Auditor General of Québec stated that other modifications should be considered to enhance the clarity and rigour of Quebec public infrastructure projects, notably as it was highlighted that the projects' governance did not necessarily favour their success (Vérificateur général du Québec 2011). Another independent report supported the one by the Auditor General and went even further in the analysis and recommendations (SECOR-KPMG 2012).

In 2013, a law on public infrastructures (*Loi sur les infrastructures publiques*) was adopted by the government, which had two major implications: the creation of a new organization, *Société québécoise des infrastructures* (SQI), and allowing SCT to produce directives in order to legally frame the governance of major infrastructure projects (Secrétariat du Conseil du trésor 2013). In 2014, SCT issued a revised framework, the '*Directive sur la gestion des projets majeurs d'infrastructure publique*' (Secrétariat du Conseil du trésor 2014). This governance framework has been revised in 2025 (Secrétariat du Conseil du trésor 2025), allowing greater flexibility, particularly regarding infrastructure maintenance projects and alternative contractual modes.

11.2 The Stage-Gate Model with Phases and Decision Points

Figure 11.1 presents the current Quebec Directive. The model applies in principle to all major government-funded projects across sectors. The fundamental steps of this model are in the front end, specifically to obtain approval from the Council of Ministers (the Executive Council) before initiating the execution phase. The whole project process is covered, yet project implementation and transfer to the operational phase are excluded. With the revised version of the framework, some projects might be approved on a case-by-case basis, notably for alternative contractual modes. Also, maintenance projects can now go from the pre-project to the planning phase, and the required documentation has been downsized to a reduced business case.

The model entails that projects be planned and executed according to the phase division shown in the figure: the pre-project, initiating phase, planning phase, execution phase,

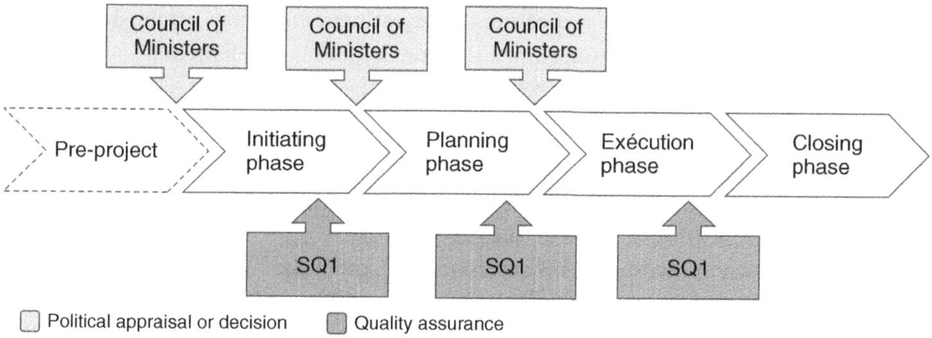

Figure 11.1 The Quebec Directive.

and closing phase. For the first three phases, important deliverables must be presented to decision-makers (the Council of Ministers) before going through the gate and being officially approved to go through the next project phase.

11.2.1 Pre-project

At this initial step, based on the various investment envelopes established by the SCT, each minister determines the major projects that will be submitted for approval for inclusion in the next *Plan québécois des infrastructure* (the portfolio, see the point on Portfolio management below) under the 'Projects under study' category.

The minister responsible for initiating a prioritized project for the public body submits a request for authorization to study the project. This request is accompanied by a Pre-Project Document (*Fiche d'avant-projet*) containing the following information:

1) A brief description of the need for the proposed project.
2) A summary demonstration that only a public infrastructure solution can meet the need.
3) A preliminary estimate of the total cost of the proposed project and the financing strategy.
4) The estimated total cost to produce each of the elements required to prepare the business case.

11.2.2 Initiating Phase

Once the pre-project has been completed, the project implementation agency designated (according to the law on public infrastructures) carries out activities related to the management and control of the project.

Following authorization from the Council of Ministers to study the proposed project, the project implementation agency must prepare an Opportunity Case (*Dossier d'opportunité*). A project subject to the environmental impact assessment and review procedure provided for in the Environmental Quality Act does not have to be the subject of an Opportunity Case, provided that the Government issues an authorization under that procedure. Infrastructure maintenance projects do not have to comply with this step neither, they go directly from pre-project to planning. This deliverable must make it possible to assess the project's relevance and recommend the best long-term option, from among those evaluated, to meet the expressed need, while respecting the investment envelopes established by the SCT. To this end, the Opportunity Case must include the following elements:

1) A needs analysis, including a demonstration that only a public infrastructure solution can respond to it.
2) Determination of the project requirements.
3) Identification and evaluation of possible options for meeting the long-term need.
4) Selection and justification of the best long-term option.
5) Annual breakdown of investments required to implement the recommended option.
6) Presentation of the main economic and financial variables.
7) The estimated cost of the elements required to prepare the Business Case for the recommended option.

8) If a PPP is chosen as the implementation method, the qualitative and quantitative arguments in support of this choice.

Once this document is authorized, the project is included in the 'Projects in planning' category of the portfolio. The amount then entered for the project corresponds to the government's investment in the estimated total cost of the option selected following approval of the project.

11.2.3 Planning Phase

The Business Case (*Dossier d'affaires*) presents, in detail, the best long-term option approved by the Council of Ministers, as well as the project management plan setting out the actions required to complete the project. It must include the following elements:

1) Project context, containing a description of the need, project requirements, and the chosen option.
2) Project scope, including project breakdown structure, analysis of functional and technical requirements, as well as the process envisaged for managing changes.
3) Estimated total project cost, its projected incremental budget impact, and financing strategy.
4) Project schedule.
5) Human resources, including a presentation of the project team and the roles and responsibilities of each member.
6) The governance structure of the project.
7) Analysis of the project's stakeholders.
8) Project risk analysis, including the probability of occurrence, potential financial impact and occurrence, its potential financial impact, and the mitigation measures envisaged.
9) Communications plan.
10) Procurement strategies, taking into account the mode of implementation envisaged.
11) Project management plan.

For infrastructure maintenance projects, only a lightweight business case has to be provided since the revised framework of 2025 has been issued, which takes into account the previous listed points but with less details.

Following this authorization, the project is included in the 'Projects in progress' category in the portfolio. The amount entered for the project corresponds to the government's maximum investment for this project, as presented in the Business Case.

Any increase in the total cost of the project presented in the approved business case constitutes a cost overrun, whether or not it increases government investment.

11.2.4 Execution Phase

Following authorization by the Council of Ministers to carry out the project, the project manager proceeds with its execution. Project progress reports must be given to the Quebec Treasury Board (SCT) yearly for the execution phase.

During this phase, any significant modification to the project must be authorized by the Council of Ministers. If the project costs increase by 10–25%, the project manager needs an authorization from SCT before proceeding. If the project costs increase by more than 25%, the project manager needs an authorization from the Council of Ministers to proceed further.

A modification of any kind is considered significant when it has the effect of modifying any of the following parameters of the Business Case:

1) The total cost of the project.
2) The project's financing strategy.
3) The scope of the project is such that it can no longer be used as a baseline.
4) The project schedule is such that it can no longer be used as a baseline.

If necessary, a modification is made in the portfolio.

11.2.5 Closing Phase

When formal acceptance of the public infrastructure is attested in writing, the minister must confirm the date of this acceptance in writing to the SCT, and the project manager must produce the project closure report.

11.2.6 Projects That Must Adhere to the Model

The Directive applies to infrastructure projects costing over $50 million (CAD). An exception has been made for road projects as they are sometimes less complex; in that case, the threshold is that projects costing over $100 million (CAD) must adhere to the Directive. For projects that fall under an alternative contractual mode (design build, integrated project delivery, etc.), the specific stage-gate to be used and documents to provide are requested on a case-by-case basis. This is only the case for projects that are initiated in the province of Quebec and under its jurisdiction, as another governance framework is required for federal projects (see Chapter 10 of this book).

IT projects are exempt from this framework, as another legislation covers them specifically, the Law on the governance and management of information resources of public bodies and government enterprises (Secrétariat du Conseil du trésor 2024a).

11.2.7 Parties and Roles

The parties and roles involved in the Quebec governance framework are presented in Figure 11.2.

- The Minister sponsoring a project (the client) approves all main deliverables before they are presented at the Council of Ministers. In case a project is initiated by a government agency or another public organization, the Ministry that has authority on it is responsible. These organizations have to contribute to the production of project documentation and to provide access to project staff and internal documents.
- The Council of Ministers approves those deliverables before the project team has authorization to start the execution phase and get access to the budgeted costs. This is the

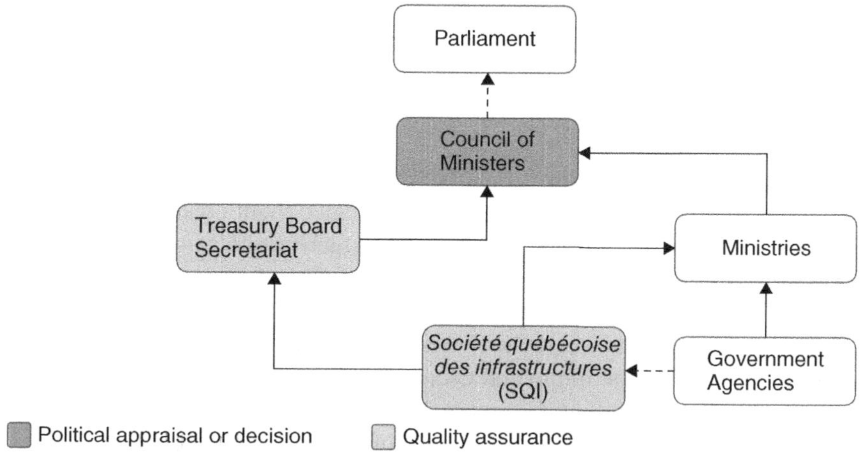

Figure 11.2 Parties and roles involved in the Quebec governance framework.

higher political authorization needed, there is no involvement of the parliament, as in other country cases.

- The SQI is the project implementation agency for most projects once the pre-project has been approved. Exceptions exists, but in the case that a government agency or another public organization request that they manage their project, SQI overviews the whole process and must accompany them. For transportation projects, the Ministry of Transport manages its own projects, without the implication of SQI, as they have their own expertise internally.
- The SCT is responsible for the governance framework of its evolution and applicability. A specific unit (Under-secretariat for public infrastructure) performs quality assurance and reviews internally the project documentation, validate the information and, emit a formal recommendation for the Council of Ministers.

11.2.8 Selected Topics

11.2.8.1 How Projects Are Initiated

Since the 2014 iteration of the governance framework, there has been a broader perspective on benefits and a more thorough evaluation of the need and justification in the front end (Samset et al. 2016). However, since every ministry and government agency can suggest their projects, there are currently many projects included in the government portfolio, too many, according to some researchers (Michaud et al. 2023). The governmental dashboard presents infrastructure projects over $20 million. So far, there are 287 projects in the planning phase and 277 in execution, from schools' renovations to the museum redevelopments. Moreover, the Quebec political party in place (Coalition Avenir Quebec) has initiated an important partisan project without going through this process (the third link between Quebec City and Levis, which is an underwater tunnel), which has brought much controversy around in the past years, especially since there is no real need behind this environmentally damaging project (Dubé et al. 2021).

11.2.8.2 Option Appraisal

In the Initiating phase, the project manager must conduct a preliminary assessment of both possible and most likely options. In the Opportunity Case, the evaluation of the most likely options for meeting the need is presented. The detailed assessment must also include the maintenance of the status quo.

The new directive adopted in 2025 is primarily intended to allow flexibility in diversifying the procurement strategies of public bodies and to give them greater flexibility in carrying out their infrastructure projects. While this is largely positive, we would like to raise an important point for consideration. While removing Opportunity Cases can potentially save a few years on timelines, the essence of the initiating phase is to evaluate all possible options. Now, a project subject to the environmental impact assessment and review procedure provided for in the Environmental Quality Act does not have to be the subject of an Opportunity Case, provided that the government issues an authorization under that procedure. We see potential risks here, even if an environmental assessment procedure is carried out, of analysing a potential solution that is not optimal for meeting the underlying needs of the project.

For each of the most plausible options, taking into account their life cycle, the detailed assessment covers the following aspects (Secrétariat du Conseil du trésor 2011):

- Technical and technological feasibility (preparatory studies).
- Impact on human resources.
- Regulatory, legal, and environmental impact.
- Consideration of sustainable development principles.
- The option's impact on the natural and built environment, including historical, heritage, and archaeological potential.
- Risk analysis, with the potential impact for each, as well as planned mitigation measures.
- Financial and sensitivity analyses.
- Benefit–cost analysis.

The purpose of the option appraisal is to determine which of the most likely options will be retained for future studies (the 'preferred option') and on the basis of which a functional and technical programme and/or a preliminary concept will be developed. The preferred option will also be the subject of an analysis of the potential construction mode of delivery.

11.2.8.3 Cost Estimation and Budget Management

At the initiating phase, while the Opportunity Case is elaborated, the contingencies included in the estimated costs range from 15% to 30%. At the planning phase, the contingencies included in the Business Case should range from 0% to 5%. This overall contingency is added to the risk reserve. As previously mentioned, during project execution, if the project costs increase by 10–25%, the project manager needs an authorization from SCT to proceed further. If the project costs increase by more than 25%, the project manager needs an authorization from the Council of Ministers to proceed further.

11.2.8.4 Quality Assurance of Decision Documents

While the 2010 version of the Quebec governance framework hired private experts to perform quality assurance, based on the Norwegian model, in the 2014 revision, their role vanished, replaced by public internal expertise. SQI is responsible for the production of project documentation. Yet, a separate internal unit is also in charge of validating the documents for quality assurance (Expertise and Project Support). SCT reviews all documentation before producing a recommendation to be submitted to the Council of Ministers along with the project documents.

11.2.8.5 Follow-Up and Ex post Evaluation

While project ex post evaluations are done internally at SQI, these are not public. The only exceptions are with regard to ad hoc reports from the Auditor General of Quebec. By reporting periodically about the state of the public projects and the institutionalized governance framework, it forces political authorities to assess the situation and to take measures to correct perceived weaknesses.

As the Quebec framework has been used for a decade now, an internal evaluation at SCT was conducted in 2021–2022 to assess what works and what should be changed. Although a local group of project management researchers was consulted as part of this assessment and shared some observations in a virtual meeting, there is no publicly available written evidence of the entire evaluation.

11.2.8.6 Guidance, Training, and Exchange of Experiences

Along with the previous version of the governance framework, the government created several training programmes in project management, coordinated by the *Centre de services partagés du Québec* (a public organization that was dismantled in 2019), and offered by universities. Some programmes aimed to develop the competencies of public project managers, while others targeted the executives involved in projects and project approvals. While these programmes have been terminated, there are specific trainings that are accessible to project managers and civil servants, but none at the institutionalized scale, which we can find in the United Kingdom, for example.

11.2.8.7 Portfolio Management

Since 2013, Quebec has been producing a 10-year infrastructure investment plan, known as the Quebec Infrastructure Plan (PQI, *Plan québécois des infrastructures*). The PQI provides data on two variables: planned investments over 10 years and a measure of infrastructure condition, the Asset Maintenance Deficit. The PQI 2024–2034 is set at an unprecedented level of $153 billion. This represents a $3 billion increase over the previous plan and a $52.6 billion increase since the 2018–2028 PQI, or 52% over a six-year period (Newswire 2024).

The proportion of investments prioritized for the long-term viability of the infrastructure assets maintenance deficit is 62%, while 38% of those investments are reserved for the construction of new infrastructure, including the construction of hospitals, seniors' homes, new schools, social housing, as well as public transit development (Secrétariat du Conseil du trésor 2024b). However, a recent report produced by independent researchers demonstrated that the current trend in infrastructure investment by the Quebec government is not sustainable (Michaud et al. 2023).

11.3 Experience with the Model

Overall, there are some positive accounts since the implementation of the governance framework. We report here the main benefits as argued by Brunet (2021):

- At the institutional level, it increases the legitimacy and credibility of the government.
- At the organizational level, it strengthens accountability by having a formal organization in charge of project management (SQI), and it develops governmental project management capabilities.
- At the project level, having to document a project thoroughly helps project teams to be more efficient.

On the side of potential improvements, here are the most important ones as highlighted by civil servants involved in those projects (Brunet 2021):

- At the institutional level, the governance framework needs more flexibility. For example, some urgent projects do not need an extensive Opportunity Case; they need to proceed as fast as possible, for example, the Maisonneuve-Rosemont hospital complex, which was so obsolete that it became a critical matter. Also, the framework is seen as restrictive, given that several delivery modes, including more collaborative alternatives such as Integrated Project Delivery. A more flexible framework would be needed to be adjusted to the mode selected, to be more aligned with the requirements, which differ from traditional delivery modes.
- At the organizational level, the teams mentioned that they require more autonomy. The governance framework clearly establishes certain roles (e.g. project implementation agency), yet it does not provide the necessary resources and means to fulfil these roles. For example, SQI, as a project implementation agency, is often very reliant on the sponsoring ministry. Thus, any given delay on their side will impact the project's performance.
- At the project level, having to document a project is seen as a heavy involvement. Opportunity Cases and Business Cases take years to develop, and sometimes, they become obsolete before the project is approved, for example, in hospital projects where new technologies require constant updates and have broader implications on the architecture and design. Also, once project documentation is submitted for approval, there are important delays before getting the authorization to start a project, between 6 months and a year on average.

When compared to other jurisdictions, there have been many legislative changes in a short period of time in Quebec. While there was from the start a strong orientation from the government to build capability in project management, this was strengthened in 2014 when the decision was taken to perform quality assurance with government experts.

The professionals involved in project teams have learned how to use it, the different actors have developed an understanding and an appropriation of the policy's requirements. The continuous evolution of the governance framework has happened as learning provided opportunities to revise and enhance those, yet it also reduces the possibility to compare project outcomes on a similar basis.

11.3.1 Remaining Challenges

The most important challenge is a lack of legitimacy of the governance framework. The current version of the framework is soon to be revised and officially announced. The latest 2025 version of the framework allows for more flexibility regarding infrastructure maintenance projects and alternative contractual modes. In parallel, the government created a new superstructure to manage public transportation projects (*Mobilité Infra Québec*), which have been hard to govern and manage efficiently in the past years (Roy 2024). The fact that there are important delays and cost overruns in projects is often attributed to the rigidity of the governance framework, especially for those that have alternative delivery modes, which need careful adjustments to be fully aligned with required deliverables. Perhaps the recent revision of the governance framework will help in that matter.

Also, some projects are still overly political, and some are not evaluated against the evidence-based facts (Dubé et al. 2021). There is a significant lack of transparency that severely hinders our ability to understand project performance and independently evaluate the overall scheme.

But most importantly, the government has tested other governance schemes with other para-public actors, and they intend to build on those experiences with the new agency *Mobilité Infra Québec* and to amend certain provisions respecting public transit. While it is interesting to inquire about why public transit projects are treated differently from other governmental projects, these policy developments set a serious precedent for other projects to come.

References

Brunet, M. (2019). Governance-as-practice for major public infrastructure projects: a case of multilevel project governing. *International Journal of Project Management* 37 (2): 283–297. https://doi.org/10.1016/j.ijproman.2018.02.007.

Brunet, M. (2021). Making sense of a governance framework for megaprojects: the challenge of finding equilibrium. *International Journal of Project Management* 39 (4): 406–416. https://doi.org/10.1016/j.ijproman.2020.09.001.

Castonguay, J. and Miller, R. (2006). La gouvernance des grands projets d'infrastructure publique: Benchmarking des cadres de gouvernance. 2006RP-20.

Dubé, J., Mercier, J., and Scanu, E. (2021). Comment survivre aux controverses sur le transport à Québec? Septentrion.

Michaud, P.C., Lévesque, L., Joanis, M. et al. (2023). Plan Québécois des Infrastructures: Comparaison interprovinciale et soutenabilité. 2023RP-28: CIRANO.

Miller, R. and Lessard, D. (2000). *The Strategic Management of Large Engineering Projects: Shaping Institutions, Risks, and Governance*. Massachusetts Institute of Technology.

Miller, R., Castonguay, J., Chebil, F., and Roy, L. (2006a). La gouvernance des grands projets d'infrastructure publique: Le Bureau des grandes infrastructures publiques. 2006RP-13: CIRANO.
Miller, R., Castonguay, J., Lareau, D. et al. (2006b). La gouvernance des grands projets d'infrastructure publique: Diagnostic et recommandations. 2006RP-22: CIRANO.
Newswire (2024). Plan québécois des infrastructures 2024-2034 - Des investissements sans précédent de 153 G$ pour les infrastructures publiques du Québec. https://www.newswire.ca/fr/news-releases/plan-quebecois-des-infrastructures-2024-2034-des-investissements-sans-precedent-de-153-g-pour-les-infrastructures-publiques-du-quebec-800461770.html (accessed 17 September 2024).
Roy, J. (2024). Coût des projets de transport collectif au Québec : peut-on faire mieux? Centre sur la productivité et la prospérité, HEC Montréal, 28 p.
Samset, K.F., Volden, G.H., Olsson, N., and Kvalheim, E.V. (2016). Governance of major public investment projects: a comparative study of principles and practices in six countries. Concept report no. 47, 137 p.
SECOR-KPMG (2012). Étude sur la gestion actuelle du plan québécois des infrastructures et sur le processus de planification des projets - Document principal.
Secrétariat du Conseil du trésor (2008). Politique-cadre sur la gouvernance des grands projets d'infrastructure publique. Québec, Canada.
Secrétariat du Conseil du trésor (2010). Politique-cadre sur la gouvernance des grands projets d'infrastructure publique. Québec, Canada.
Secrétariat du Conseil du trésor (2011). Guide d'élaboration du Dossier d'Affaires des grands projets d'infrastructure publique. Québec, Canada.
Secrétariat du Conseil du trésor (2013). Loi sur les infrastructures publiques. Québec, Canada.
Secrétariat du Conseil du trésor (2014). Directive sur la gestion des projets majeurs d'infrastructure publique. Québec, Canada.
Secrétariat du Conseil du trésor (2024a). Loi sur la gouvernance et la gestion des ressources informationnelles des organismes publics et des entreprises du gouvernement. Québec, Canada.
Secrétariat du Conseil du trésor (2024b). Le Plan québécois des infrastructures 2024–2034. Québec, Canada.
Secrétariat du Conseil du trésor (2025). Directive sur la gestion des projets majeurs d'infrastructure publique. Québec, Canada.
Vérificateur général du Québec (2004). Rapport du Vérificateur général du Québec à l'Assemblée nationale pour l'année 2003–2004: Annexe A – Rapport de vérification concernant la gestion du projet de prolongement du réseau de métro sur le territoire de la Ville de Laval.
Vérificateur général du Québec (2011). Rapport du Vérificateur général du Québec à l'Assemblée nationale pour l'année 2011–2012: Vérification de l'optimisation des ressources, ch 1: Infrastructures p. 1.1–1.15.

12

Australia

Dominic Ahiaga-Dagbui[1] *and Jon Frazer*[2]

[1] *School of Architecture and Built Environment, Deakin University, Geelong, Australia*
[2] *Independent Infrastructure Consultant, Port Macquarie (NSW), Australia*

Australia's infrastructure governance landscape is shaped by its federal structure, vast geography, and growing pipeline of large-scale public investments. While the Federal Government has established a national Gateway Review Process to guide project planning and delivery, each state and territory operates its own frameworks, often tailored to local priorities and risk thresholds. This multi-layered system aims to improve transparency, accountability, and value for money but is frequently challenged by political influence, fiscal imbalances, and inconsistent compliance. This chapter outlines the structure and operation of Australia's governance models, highlighting the role of assurance reviews (ARs), strategic planning, and institutional reforms in shaping infrastructure outcomes.

12.1 Context for the Governance Model

Australia's history of infrastructure decision-making can be broadly characterized by scale and ambition. Projects are often large, costly, or risky. Some of this is by necessity – with one of the lowest population densities in the world and more than 1000 km between its three largest cities, Sydney, Melbourne, and Brisbane, project economics is often stretched by distance. Australia's system of infrastructure decision-making is shaped by its Constitution, which in 1901 divided the nation's previously British colonies into six semi-sovereign State Governments under a Commonwealth Government (Federal Government from here onwards). Two Territory Governments were later established at the state level. The Federal Government retains control over areas of national interest such as telecommunications, aviation, and infrastructure regulation, while transport, health, education, and most other forms of service are delivered by state governments.

Since federation, project funding has remained a point of contention. While states are responsible for the bulk of infrastructure delivery, the Federal Government retains the greatest power to raise revenue through income tax collection. The result is a vertical fiscal imbalance, whereby states rely heavily on federal transfers to pay for infrastructure investments, services, and maintenance, and the Federal Government seeks to influence how

Governance Frameworks for Major Public Projects: International Practices and Experiences, First Edition.
Edited by Vedran Zerjav, Morten Welde and Gro Holst Volden.

and where the funding is applied. This remains the most fundamental fault line underlying the nature, process, and outcomes of project governance in Australia. For example, governments' competing interests and overlapping electoral cycles often mean that political interests drive many infrastructure decisions. This politicization of infrastructure has led to commitments to megaprojects, both formally and de facto, before any significant analysis has been undertaken.

The stakes in project governance have risen, with Australia's public infrastructure pipeline growing significantly in recent years. This is boosted by major investments in public transport, telecommunications, and energy, with a total budget of A$270 billion over the four years to FY2027–28 (Infrastructure Partnerships Australia 2024). The full pipeline of major projects in Australia totals in excess of A$1 trillion, with approximately A$700 billion in the planning and development phases (Infrastructure Partnerships Australia 2025). Projects have also grown rapidly in scale, with 15 projects with a total estimated capital expenditure of over A$5 billion currently under construction, only 15 years after Australia's first project of this magnitude.

12.2 Project Governance Models for Infrastructure Investments

The Federal Government's Gateway Review Process, as implemented by the Department of Finance, serves as the overarching governance framework for infrastructure projects across the nation (Department of Finance 2016). The Federal Government adopted the Gateway Review Process in its 2006–2007 Budget as part of a broader effort to improve public sector efficiency and project delivery outcomes. This model, adapted from the UK model, was introduced in response to the need for a systematic and independent assessment process to ensure that major projects are adequately planned, appraised, and delivered with transparency and accountability. ARs are conducted for procurement or infrastructure projects with a total estimated cost of A$30+ million, A$10+ for ICT projects and programmes with a total estimated cost of A$50+ million (Department of Finance 2016). Defence projects are not assessed using the Gateway review process outlined in this chapter – these are assessed by Cabinet under a separate model.

12.2.1 Structure of the Federal Government's Gateway Review Process

The Federal Government's Gateway Review Process has six stages, from establishing a business need to ex post evaluation. These stages, the ARs at each gate and Decision Points (DPs), are outlined in Figure 12.1 and include:

- **Business need:** Assures that the scope and purpose have been adequately assessed, the project fits within the entity and/or the government's strategies and policies, and the expected benefits have been identified.
- **Business case:** Focuses on the project's objectives, costs, benefits, risks, and delivery options.

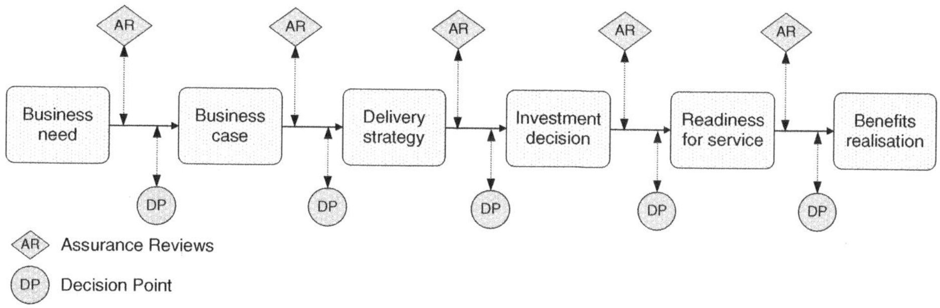

Figure 12.1 Federal government's gateway review process.

- **Delivery strategy:** This ensures that the procurement strategy clearly defines the project's scope, benefits, and plan for implementation and whether the project is ready to enter procurement.
- **Investment decision:** Ensures that the supplier selection process is robust and aligned to the projects' scope and purpose and confirms that the project is ready for contracted delivery, with benefit management strategies and plans incorporated.
- **Readiness for service:** Ensures that the solution is robust before delivery, assesses organizational readiness both before and after delivery, and establishes the basis for evaluating ongoing performance.
- **Benefits realization:** Assesses whether the project has delivered the anticipated benefits and actual costs aligned with the projected budget and whether lessons can be learned to improve future projects. This is typically undertaken between 6 and 24 months of project completion, depending on the project type.

12.2.2 Assurance Reviews

ARs at the gates (or simply, reviews) are conducted by independent assessors from consulting firms, specialized agencies, government departments, or infrastructure advisory bodies such as Infrastructure Australia or Infrastructure Victoria. Reviews are short and focused – typically lasting three to five days, during which the team conducts interviews and evaluates documentation, providing an objective assessment of the project at each gate, identifying risks, and making recommendations for corrective actions.

The Project Steering Committee or Senior Responsible Officer (SRO) is usually the ultimate decision-maker at each gate. The SRO is typically a high-ranking official within the organization or government body sponsoring the project. This role is often filled by someone like a Director-General, Minister, Chief Executive Officer, or Secretary. For particularly large or complex projects, the SRO might be someone with substantial project delivery experience or expertise in the relevant sector, such as transportation, health, or education. The SRO is responsible for ensuring that any necessary corrective actions are taken after receiving feedback from the assurance reviewers. The recommendations of the reviews are largely advisory, and SROs can still proceed with a project, even if there is a strong recommendation not to do so.

12.2.3 State-Based Gateway Reviews and Special Pathways

While the Federal Government has established its own set of guidelines under the national Gateway Review process, each state and territory has developed its own versions of the process, tailored to meet local legislative, economic, and social requirements. As shown in Figure 12.2, Australia employs a wide range of infrastructure investment frameworks depending on the state, level of risk associated with a project, and level of expected capital investment. For example, the State of Victoria operates two frameworks – the Gateway Review Process for medium-risk infrastructure and information and communication technology (ICT) projects, and the High-Value High-Risk (HVHR) framework,[1] which imposes additional scrutiny on projects that exceed A$100 million or are deemed particularly risky.

The diverse approaches adopted by various frameworks can create conflicting requirements for projects that span multiple jurisdictions or are subject to different governance layers. For example, in addition to state-based ARs, the so-called 'nationally significant' proposals,[2] which request more than A$250 million from the Federal Government, also have to go through the Four-Stage Assessment Framework administered by Infrastructure Australia (Infrastructure Australia 2021). Operating under multiple frameworks increases the financial and human resource burden on project proponents. Each framework requires specific documentation, resulting in high planning and administrative task costs.

12.2.4 Flexibility, Exceptions, Special Programmes, and the Gateway Review Model

While the landscape of project governance in Australia may appear bureaucratic and complex, some flexibility is embedded in the framework. This flexibility is crucial for accommodating the range of projects that fall under this purview, particularly given the varying scales, risks, and strategic importance of these projects across Australia. Some projects bypass the stages of the Gateway Review Process or follow alternative paths. Special government programmes such as the National Water Infrastructure Fund, Urban Congestion Fund, Black Spot Program, and Commuter Car Park Fund have their own streamlined criteria. These projects might avoid full gateway scrutiny, depending on ministerial oversight and specific guidelines of the funding programme.

---

1 According to the Victorian Department of Finance, projects are High-Value High-Risk if:

- considered high risk when assessed as such using the Department risk assessment tool, the Project Profile Model (PPM).
- considered medium risk using the PPM and has a Total Estimated Investment (TEI) of between A$100 and A$250 million.
- considered low risk using the PPM but has a TEI over A$250 million.
- identified by the Government as warranting the rigour applied to HVHR investments.

2 Infrastructure Australia is the nation's independent infrastructure advisor, responsible for evaluating proposals for investment in nationally significant infrastructure and other infrastructure determined by our responsible Minister. Nationally significant proposals are projects requesting more than $250 million in Commonwealth Government funding.

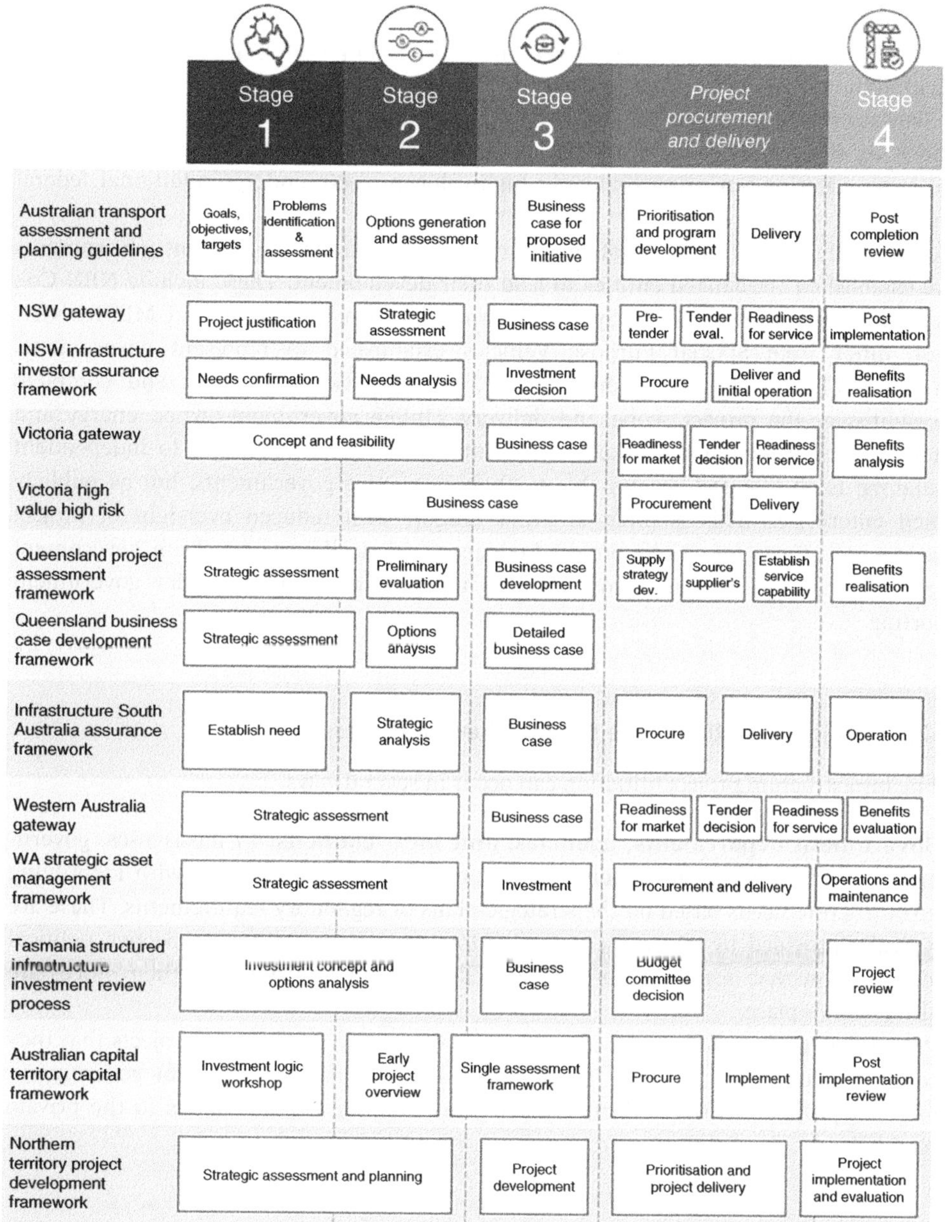

Figure 12.2 State and territory governance frameworks aligned with Infrastructure Australia's Four-Stage Assessment Framework. *Source:* Infrastructure Australia (2021) / Infrastructure Australia / CC BY 3.0 AU.

Another element of flexibility in the governance process is its application based on capital thresholds and the significance of the project. For example, only projects that request A$250 million or more in funding from the Federal Government have to go through the gateway review administered by Infrastructure Australia (Infrastructure Australia 2021). Otherwise, the review is only undertaken at the state level. This threshold serves as a filtering criterion, ensuring that only high-value projects undergo additional federal review.

Additionally, as the number of megaprojects in Australia has grown, governments have established specialized entities to lead their development. These include NBN Co., Inland Rail and Snowy Hydro, Sydney Metro, Suburban Rail Loop, and METRONET. These differ from Special Purpose Vehicles established to represent governments through public–private partnerships (PPPs), which have more direct and complete oversight over the project scope and delivery. Unlike government-owned energy and water utilities in regulated markets, these entities are not accountable to independent regulators. Each remains accountable to their respective governments, but as publicly owned enterprises. This is often at arm's length, with reduced oversight of project decisions and financing. Their project budgets are typically separate from government balance sheets, meaning that cost changes are not reflected in regular government reporting.

12.2.5 Project Initiation in the Gateway Review Process

Public infrastructure project initiation can occur in several ways.

- **Government departments, agencies, and local councils:** In most cases, governments of all levels through their departments or entities are tasked with identifying infrastructure needs based on the strategic plans or regulatory requirements. These are typically provided by governments as 'bottom-up' advice to their respective political leaders. However, 'top-down' project suggestions from Ministers and their offices are also commonplace.
- **Private sector initiation:** Private entities may propose and initiate projects that they could develop as proponents. These market-led proposals differ from government-initiated projects, as they effectively delegate planning responsibilities to the private sector.

12.2.6 Options Analysis

All Australian governance frameworks assume that a *reasonable* range of conceptually different options is considered in the early stages of project development. For example, Infrastructure Australia's Guidance for Business case development recommends the inclusion of at least two options, in addition to a 'do-minimum' base case (Infrastructure Australia 2021). This is typically either expected as a standalone stage before Final Business Case development or incorporated into the Strategic Business Case (see Figure 12.2).

12.3 Experience with the Models

Australian governments' records of implementing best-practice models for project governance are mixed. The quality of advice has improved substantially, with governments developing more sophisticated tools, governance processes, and requirements to improve project appraisals and outcomes. However, lingering issues remain regarding political interference in the selection process, the transparency of decision-making, and the effectiveness of the AR process.

12.3.1 Improvements in the Quality of Advice on Projects

Many institutional arrangements, processes, and frameworks informing decisions about public spending on infrastructure have been greatly enhanced over the past two decades. The sophistication of advice on public infrastructure investments has improved alongside rapid growth in the scale and complexity of the investments themselves. This has enabled the delivery of urban transport projects of unprecedented scale in Australia, such as the concurrent development of several new Sydney Metro networks.

Three notable organizational improvements have yielded substantial benefits for project governance in Australia.

- **Independent infrastructure agencies:** Australia was an early adopter of independent advisory bodies, with Infrastructure Australia established in 2008 as a statutory advisory council to provide independent advice on infrastructure investments based on long-term needs rather than short-term political interests. Most State and Territory Governments have set up similar agencies to advise their governments on infrastructure issues, including Infrastructure New South Wales in 2011, Infrastructure Victoria and Infrastructure Tasmania (2015), Infrastructure South Australia (2018), Infrastructure Western Australia (2019), and Infrastructure Northern Territory (2021). All these agencies share a common motivation: to enhance the quality of advice to governments and to provide an expert voice on infrastructure within decision-making apparatuses.
- **Gateway review processes:** Established in each State and Territory as well as at the Federal Government level, these processes have helped to support evidence- and needs-based evaluation of projects, with independent and multi-disciplinary review of the economic, environmental, and social impacts of each proposal.
- **HVHR and Project Assurance frameworks:** Complementing the Gateway Reviews, the HVHR frameworks introduced by various governments have added a layer of scrutiny and transparency to complex or capital-intensive projects. Some States also utilize a Project Assurance Review framework to ensure that projects in the delivery stage remain 'on track'. This recognizes that complex projects must always deal with unforeseen issues and uncertainties during their delivery.

These processes have undoubtedly enhanced the capacity of Australian governments to filter projects with relatively weak economic, environmental, and social credentials at an earlier stage of development.

Vignette: City to Gungahlin Light Rail, Canberra

A success story of intergovernmental gateway reviews is the City to Gungahlin Light Rail project in Australia's capital city, Canberra. In 2012, the Australian Capital Territory (ACT) Government submitted an initial project proposal to Infrastructure Australia with two options for a 12-kilometre public transport link. These were a light rail, with an initial benefit–cost ratio of 1.02, and bus rapid transit, which yielded an estimated ratio of 1.98 (ACT Government 2012). After conducting further analysis of the social impacts of each option, the ACT Government opted to proceed with the light rail option in 2013 on the basis that it would deliver 'the best overall outcome for Canberra' (Corbell 2013). This attracted considerable criticism, with Infrastructure Australia opting against endorsing the project for federal funding. However, the ACT Government responded to the debate with transparency, publishing a subsequent business case for the preferred option in full to further justify its decision (Capital Metro Agency 2014). The project scope was refined through review, yielding an updated benefit–cost ratio of 1.2 and an estimated capital cost of A$783 million.

The agency tasked with delivery, the Capital Metro Agency, later known as Transport Canberra, chose to procure the project as a Public Private Partnership (PPP). Following procurement, the contract was published in 2016, detailing a 20-year Availability PPP with the Canberra Metro consortium and a lower construction cost of A$707 million (Capital Metro Agency 2016). The project was delivered on time, with services commencing in April 2019. It was also completed under budget at a total cost of A$675 million (Transport Canberra 2019). The ACT Government undertook significant urban renewal along the project corridor to accelerate the social benefits of the project, which it measures through a Benefits Realisation framework. The post-completion review for the project, released in 2019, was the only publicly available report of its kind for an Australian infrastructure project with a capital cost of over A$500 million at that time (Major Projects Canberra 2020).

12.3.2 The Battle Over Infrastructure Decisions: Governance, Politics, and Options Analysis

Despite the robust and multilayered processes of project governance in Australia, the tension between politics and rational evidence-based planning during project initiation is a perennial issue in infrastructure development. Most Australians believe infrastructure decision-making is highly politicized, with 59% indicating that projects are approved for political reasons and 68% saying politicians have too much influence in project selection (ANU 2022).

Bureaucrats working within the governance framework rarely make final decisions on projects; they may delay or request modifications to underdeveloped proposals through Gateway Review mechanisms. However, the final decision on the project's selection is made by SROs, typically Ministers, who have no obligation to heed evidence-based advice. A project's merits may thus form one part of Minister's decision-making, but it appears that political considerations are often more pertinent – especially for 'flagship' projects that are likely to garner public support in marginal electorates. For example, the scope of the National Broadband Network (NBN) was altered repeatedly by successive governments from its inception in 2007 to midway through its delivery a decade later, and this continues to evolve even in 2025. The project was commenced *without* a business case with an initial public investment of A$4.7 billion and a total capital cost of approximately A$43 billion, with much of that cost to be met by private investment (Martin 2010). The total cost has since increased to A$57 billion for an inferior scope (Environment Communications Legislation Committee 2020).

For many megaprojects, it appears that a project's selection and funding remains a cultural and political hurdle rather than an economic or procedural hurdle. Infrastructure Australia (2018) thus noted that

> *Too often, we see projects being committed to before a business case has been prepared, a full set of options have been considered, and rigorous analysis of a potential project's benefits and costs has been undertaken.*

For example, the 90 km Suburban Rail Loop (SRL) in Victoria was announced three months before the 2018 state elections and prior to the completion of the strategic business case (ABC News 2022). The project was initially announced by the state government at a cost of A$50 billion (Oaten 2018). Estimates from the State's Parliamentary Budget Office in 2024 indicate that the project could cost at least A$125 billion for construction alone (PBO 2024). The Auditor General noted that the 'government made funding commitments to the projects' prior to the development of a business case (VAGO 2022b).

In some circumstances, funds are ring-fenced for specific purposes, with funding eligibility requirements drafted to circumvent broader scrutiny or the need to deliver net public benefits. An example is the Commuter Car Park Projects, where the ANAO (2021) found that funding allocations were made without adequate evidence or analysis and that the process was effectively designed to deliver political outcomes. In this case, projects were selected after political offices were canvassed for suggestions, resulting in 77% of sites chosen for funding being located in electorates held by the government and a further 10% in marginal electorates targeted by the government (ANAO 2021). In another example, Snowy Hydro 2.0 was fast-tracked through political commitments as a 'nation-building' project to meet Australia's energy storage needs. Despite the project's scale and complexity, the decision to proceed with Snowy Hydro 2.0 was made quickly, largely bypassing the comprehensive exploration of other potential solutions that the Gateway Review Process is supposed to facilitate. At a revised price tag of A$12 billion, the project is expected to cost at least six times more than its original budget (Lowrey 2023).

After a critical review commissioned by the Infrastructure Minister on the planning and cost of infrastructure projects in 2023, the Federal Government cancelled funding for 50 infrastructure projects worth A$11.6 billion that were approved by the previous government (Jeffrey 2023). Some of these projects are already under delivery. The Review noted that

> *there are projects in the [infrastructure investment plan] that do not demonstrate merit, lack any national strategic rationale and do not meet the Australian Government's*

national investment priorities... A number of projects were allocated a commitment of Australian Government funding too early in their planning process and before detailed planning and credible design and costing were undertaken. (p. 4)

In politically charged environments, the desire to maintain momentum behind large-scale projects and demonstrate proactive governance typically results in the dismissal of the 'do-nothing' option early in the process. Once a project is publicly announced or endorsed by political leaders, the bureaucratic governance process appears to work in reverse, with the options analysis and subsequent appraisal stages becoming exercises in validation rather than genuine evaluation. This has led to some criticism that the governance process, particularly for large projects, merely rubber-stamps projects that have already been politically endorsed. This practice undermines the integrity of the options analysis process, as it does not genuinely explore a range of feasible alternatives but rather steers the decision towards a predetermined outcome.

12.3.3 Ex Post Evaluation as Part of the Gateway Review Process

Ex post evaluation, the assessment of whether a project has delivered its intended objectives, is a key part of the Gateway Review framework outlined in Figure 12.2. However, notable gaps exist in the completion of post-project evaluations. For example, in Victoria, all HVHR projects must complete an ex post evaluation (Gateway 6) 6–18 months after project completion. However, as can be seen in Table 12.1, ex post completion rates since 2013 have been dismal. This low completion rate can be attributed to several factors, including budget constraints, shifting priorities once a project is operational, and the complex nature of aligning final project outcomes with initial projections. Some entities cited insufficient budget allocations for conducting a thorough Gate 6 review or ongoing work in earlier gates, which delayed the completion of the ex post evaluation.

12.3.4 Effectiveness of Assurance Review

The Gateway Review model relies on sponsoring agencies and SROs to act on the recommendations provided by the reviewers. However, it is common for agencies to ignore or only partially accept review recommendations. For example, on the Ultranet e-learning platform project, reviews flagged many critical risks. However, the project forged ahead without fixing the issues. The responsible department failed to adequately address over 60% of the review recommendations (VAGO 2013). The project faced significant challenges, including design and usability flaws, ultimately becoming an expensive failure that cost 80% more than its initial A$60 million budget. It was delivered six years later and discontinued three years after its launch (ANAO 2012).

Thus, the AR provides a framework for valuable independent feedback, but its effectiveness is blunted if agencies do not act on the findings and recommendations. It can help identify critical risks, but cannot force mitigation – that responsibility rests with the SRO. Ignoring review findings – often for political or bureaucratic convenience – negates the intended benefits of ARs.

Table 12.1 Ex post evaluation (Gateway 6 completion) in the State of Victoria.

Year (Source)	Report title	Completion of gateway 6 (ex post evaluation)	Key insights
2013 (VAGO 2013)	Planning, Delivery and Benefits Realisation of Major Asset Investment: The Gateway Review Process	No projects completed Gateway 6	No single project had completed the full suite of Gateway reviews since the model's introduction in 2003. Nearly 70% of projects reviewed had completed two or fewer gates, with none reaching Gateway 6.
2022 (VAGO 2022a)	Major Projects Performance Report 2022	Only 2/24 HVHR projects completed Gateway 6	Ex post evaluation completion is still very low. Budget constraints and shifting priorities were cited as reasons for the low completion rate.
2023 (VAGO 2023)	Major Projects Performance Report 2023	Only 2/25 HVHR projects completed Gateway 6	In 2023, while more projects had progressed through earlier stages, the completion of Gateway 6 remained low. Several projects either delayed their post-completion reviews or had not yet undergone the Gate 6 assessment.

Furthermore, the effectiveness of gateway reviews depends on honest and evidence-based recommendations by the reviewers. There is evidence suggesting that some review agencies, particularly 'so-called' independent government infrastructure advisory bodies like Infrastructure Australia, may be hesitant to be overly critical of proposals, especially when the government has invested significant political or financial capital in their development. One example was the Inland Rail project – a 1700-km freight rail connection between Melbourne and Brisbane – which failed to quantify the relative costs and benefits of alternative options, such as upgrading the existing interstate highway. In its evaluation of the business case, Infrastructure Australia called for a more complete, transparent, and objective assessment of the options considered and was critical of several assumptions underpinning an already marginal economic case for the project (Infrastructure Australia 2016). Despite these concerns, Infrastructure Australia endorsed the project anyway, and it proceeded in the form previously committed by the Federal Government. The estimated cost of the project has since increased from A$4.7 billion in 2010 to A$31.4 billion in 2022 (Schott 2023). The estimated construction completion date has stretched from 2025 to 2031, with some sections 'on hold' (Chan et al. 2024).

12.3.5 Poor Transparency

While there is broad consensus on the value of transparency and the review of public infrastructure decisions within Australia's democratic system, governments rarely release the advice used to make these decisions. By design, gateway reports are confidential and typically only shared with the SRO and Assurance Review Unit. Only summary statistics or lessons learned are occasionally published in aggregate. Thus, most of the advice informing ministers' infrastructure decisions is – and will remain – in-confidence. This lack of public transparency makes it nearly impossible to accurately measure government success in preventing relatively weak or potentially risky projects from being built. It also makes it challenging to quantify the value of investment in productive infrastructure that has occurred as a result. Transparency is enhanced by audit offices in each federal, state, and territory government, which can review past decisions, projects, and funding programmes with varying levels of independence and autonomy. Similarly, Independent Reviews appointed by subsequent governments and led by industry experts can shed light on decision-making. However, it is often too late to correct poorly informed decisions by the time these reviews are conducted.

12.4 Conclusion

The governance of infrastructure projects in Australia involves a complex interplay of political, fiscal, and bureaucratic forces. While the country has made significant strides in developing sophisticated frameworks, the persistent influence of political interests has often undermined these efforts. The establishment of independent advisory bodies and the adoption of structured review processes have contributed to more informed decision-making and strategic planning. For gateway reviews to reach their full potential, cultural, and institutional changes are needed. This includes mandating genuine compliance (no opt-outs) with all gates for high-risk projects, improving public transparency (e.g. publishing business case summaries or gateway outcomes to the extent possible), and building political norms that prize long-term value over short-term expediency. The model works best when all parties – project agencies, oversight bodies, and political decision-makers – buy into the ethos of independent scrutiny, honest disclosures, follow-through, and continuous improvement. Resolving the tension between political and bureaucratic forces is essential to achieving best-practice project governance in Australia. This will help Australia to build a robust and sustainable infrastructure legacy that meets the needs of its citizens and supports the nation's growth and development.

Acknowledgements

The authors acknowledge the use of generative AI tools Copilot and ChatGPT in surfacing insights from large and multiple government reports, business cases, and Auditor General reports. Paperpal was used in copy-editing of the text to improve the overall quality of the content. The authors carefully reviewed any AI-generated output for accuracy and coherence. The final content, conclusions, and assertions in this chapter are our own.

References

ABC News (2022). Independent analysis puts the cost of building and running Victoria's Suburban Rail Loop at $200 billion. *ABC News*. https://www.abc.net.au/news/2022-08-18/budget-analysis-puts-rail-line-project-at-200-billion-dollars/101344828 (accessed 21 September 2024).

ACT Government (2012). City to Gungahlin transit corridor: city to Gungahlin transit corridor, infrastructure Australia, accessed 24 February 2025.

ANAO (2012). Administration of the gateway review process, Department of Finance and Deregulation, Audit Report No. 22 2011–12, Australian National Audit Office (ANAO), accessed 12 October 2024.

ANAO (2021). Administration of commuter car park projects within the Urban Congestion Fund Australian National Audit Office (ANAO), accessed 31 October 2024.

ANU (2022). Trust in government key to Australia's big build. Australian National University, *ANU News*. https://www.anu.edu.au/news/all-news/trust-in-government-key-to-australias-big-build (accessed 29 March 2025).

Capital Metro Agency (2014). Capital metro: full business case. ACT Government. https://www.cityservices.act.gov.au/__data/assets/pdf_file/0010/887680/Light-rail-Capital-Metro-Business-Case-In-Full.pdf (accessed 25 February 2025).

Capital Metro Agency (2016). Contract summary. ACT Treasury. https://www.treasury.act.gov.au/__data/assets/pdf_file/0008/1907711/Capital-Metro-Project-Summary.pdf (accessed 25 February 2025).

Chan, G., Bowers, M., Ball, A., and May, N. (2024). Trouble on the tracks: is Australia's $14bn inland rail project going off the rails?*The Guardian*, 30 October 2024. https://www.theguardian.com/australia-news/ng-interactive/2022/jan/25/trouble-on-the-tracks-is-australias-40bn-inland-rail-project-going-off-the-rails?utm_term=64405652009c480d45a248c2b8a906df&utm_campaign=TheRuralNetwork&utm_source=esp&utm_medium=Email&CMP=ruralnetwork_email.

Corbell, S. (2013). Light rail best choice for Canberra's future. *My Gungahlin*. https://mygungahlin.com.au/light-rail-best-choice-for-canberras-future/556/ (accessed 25 February 2025).

Department of Finance (2016). Australian Government Assurance Reviews, Department of Finance (Australian Commonwealth Government). Commonwealth of Australia. www.finance.gov.au/assurance-reviews/ (accessed 2 October 2024).

Environment Communications Legislation Committee (2020). Official Committee Hansard: Budget Estimates. Commonwealth of Australia, accessed 21 October 2020.

Infrastructure Australia (2016). Inland rail: project business case evaluation. Infrastructure Australia, accessed 15 October 2024.

Infrastructure Australia (2018). Infrastructure decision-making principles. Infrastructure Australia. https://www.infrastructureaustralia.gov.au/publications/infrastructure-decision-making-principles (accessed October 2024).

Infrastructure Australia (2021). Assessment framework, Australia. Infrastructure Australia. https://www.infrastructureaustralia.gov.au/develop-proposal (accessed 15 December 2025).

Infrastructure Partnerships Australia (2024). 2024 Australian infrastructure budget monitor. (accessed 13 February 2025).

Infrastructure Partnerships Australia (2025). Australia and New Zealand infrastructure pipeline. https://infrastructurepipeline.org/ (accessed 13 February 2025).

Jeffrey, D. (2023). Fifty infrastructure projects axed as minister blames former government's 'economic vandalism'. *Nine News*. https://www.9news.com.au/national/federal-politics-infrastructure-projects-axedindependent-review-minister-catherine-king/bed768bf-ca24-4704-bdd4-d5b7229ae558 (accessed October 2024).

Lowrey, T. (2023). Snowy hydro expansion hits reset button as costs blow out to $12 billion. *ABC News*. https://www.abc.net.au/news/2023-08-31/snowy-hydro-reset-project-to-cost-12-billion/102797650 (accessed September 2024).

Major Projects Canberra (2020). City to Gungahlin light rail: benefits realisation – snapshot, ACT Government, accessed 21 February 2025.

Martin, I. (2010). The promised land: costs and benefits of the NBN vision. *Telecommunication Journal of Australia* 60 (2): 30. https://doi.org/10.4225/03/5906b279869ae.

Oaten, J. (2018). Does Melbourne's $50-billion suburban rail loop risk eating up 'a generation's worth of spending'? *ABC News*. https://www.abc.net.au/news/2018-10-22/melbourne-50-billion-dollar-rail-loop-planning-questioned/10385702 (accessed 3 October 2024).

PBO (2024). Suburban rail loop: cost to build and operate East and North sections, Parliamentary Budget Office, Victorian Parliamentary Budget Office. pbo.vic.gov.au (accessed 4 October 2024).

Schott, K. (2023). The delivery of inland rail: an independent review. Department of Infrastructure, Transport, Regional Development, Communications and the Arts. https://www.infrastructure.gov.au/sites/default/files/documents/independent-review-of-inland-rail-report.pdf (accessed 4 October 2024).

Transport Canberra (2019). City to Gungahlin light rail: project delivery report. ACT Treasury, accessed 3 October 2024.

VAGO (2013). Planning, delivery and benefits realisation of major asset investment: the gateway review process, Victorian Auditor General, accessed 18 June 2024.

VAGO (2022a). Major projects performance reporting 2022, independent assurance report to Parliament 2022–23: 6, Victorian Auditor-General's Office, accessed 9 August 2024.

VAGO (2022b). Quality of major transport infrastructure project business cases, Victorian Auditor-General's Office, accessed 9 September 2024.

VAGO (2023). Major projects performance reporting 2023, independent assurance report to Parliament 2023–24: 9, Victorian Auditor General's Office, accessed 19 October 2024.

13

Governance Arrangements for EU-Funded and Nationally Funded Public Investment in the EU

*Cristiana Belu Manescu** and *Ana-Maria Dobre**

European Commission, Directorate General for Economic and Financial Affairs, Brussels, Belgium (Cristiana Belu Manescu); Directorate General for Employment, Social Affairs and Inclusion, Brussels, Belgium (Ana Maria Dobre); Ana Maria Dobre's contribution was provided while affiliated with the Directorate General for Regional and Urban Policy, Brussels, Belgium.

The governance of public investment in the European Union (EU) reflects the complexity and diversity of its 27 Member States, each with distinct institutional structures, economic profiles, and administrative traditions. While national systems manage most public investments, EU-funded projects – particularly those under Cohesion Policy – are subject to a multilevel governance framework involving the European Commission and national and regional authorities. This dual structure introduces both harmonized standards and varied practices across the investment cycle. This chapter explores the principles, mechanisms, and institutional arrangements that underpin EU-funded and nationally financed public investments, highlighting the evolving role of strategic planning, appraisal, budgeting, and oversight in promoting cohesion, efficiency, and accountability.

13.1 Introduction

The European Union (EU) is a mix of 27 mostly well-developed nations, of which 22 are members of the Organisation for Economic Co-operation and Development (OECD) and three are in the process of negotiating membership. Yet, there are significant differences in the level of economic development between these nations. The EU is also very diverse institutionally, with large and small states, unitary and federal states, which translates into the rich institutional arrangements for infrastructure governance observed in practice (Ruiz Rivadeneira and Mcmaster 2023).

The actual management of public investments in the Member States is influenced by the share of EU Cohesion Policy funding. In those Member States with a large share of Cohesion Policy in public investment, the EU rules and procedures for Cohesion Policy dominate, while the national system of public investment co-exists for a smaller share of projects. There are dedicated line ministries that oversee specific requirements

* The information and views set out in this article are those of the author and do not necessarily reflect the official opinion of the European Commission.

Governance Frameworks for Major Public Projects: International Practices and Experiences, First Edition. Edited by Vedran Zerjav, Morten Welde and Gro Holst Volden.

for planning, selection, monitoring, and control systems as required by Cohesion Policy rules. By contrast, in those Member States with a small share of EU Cohesion Policy, the management of public investment is largely unaffected by EU procedures/rules. Section 13.2 discusses the role and specifics of the EU Cohesion Policy, as established in the Treaty of the European Union, and discusses its multilevel governance framework between the European Commission and the Member States' national and regional authorities.

Other than projects financed by EU Cohesion Policy, the governance of public investment projects falls under the responsibility of the Member States. There are very limited requirements in the EU legal framework targeting national practices of managing public investment in the Member States. Existing requirements cover, for example, public procurement aspects, railway infrastructure, and data publication requirements for extra-budgetary units and funds and for contingent liabilities.[1]

Yet, the efficiency of public investment management is also a quality of public finances issue, with possible fiscal implications. It is, therefore, relevant for the EU policy framework to contain excessive deficits as enshrined in the founding provisions of the Treaty on the Functioning of the European Union (TFEU) (Article 126). The latest EU governance framework reform of 2024 puts, among others, an emphasis on medium-term planning as a tool for sound and sustainable public finances while also promoting growth through reforms and investments.[2]

This chapter is organized as follows: Section 13.2 discusses the principles and the multilevel governance framework underpinning EU Cohesion Policy; Section 13.3 reviews governance practices in the planning, appraisal, budgeting, implementation, and ex post reviews of nationally financed public investments in the EU Member States; and Section 13.4 concludes.

13.2 Governance Arrangements for EU Cohesion Policy-Funded Investments

The EU Cohesion Policy is key for promoting economic, social, and territorial cohesion among Member States. The policy supports investment for jobs and growth, through funds such as the European Regional Development Fund (ERDF), Cohesion Fund (CF), European Social Fund Plus (ESF+), and Just Transition Fund (JTF). It also supports the Interreg European Territorial Cooperation goal.

For over three decades, a significant share of the EU budget – currently €392 billion for 2021–2027 – has been specifically designated to achieve the objective enshrined in the TFEU of enhancing the economic, social, and territorial cohesion of the EU by mitigating territorial imbalances and promoting 'coordinated development' between countries and regions.

1 For example, Directive 2014/24/EU, as amended by three Commission Delegated Regulations 2015/2170, 2017/2365, 2019/1828. Directive 2012/34/EU and Directive 2011/85, as amended by Council Directive (EU) 2024/1265.

2 The 2024 European Governance Reform consists of two regulations and one Directive, namely Regulation (EU) 2024/1263 of the European Parliament and of the Council of 29 April 2024 on the effective coordination of economic policies and on multilateral budgetary surveillance and repealing Council Regulation (EC) No. 1466/97; amendments to Council Regulation (EC) No. 1467/97 of 29 April 2024; and Council Directive 2011/85/EU of 8 November 2011 as amended by Council Directive (EU) 2024/1265 of 29 April.

This calls for particular attention to regions with lagging development, such as rural areas, those undergoing industrial transition, and regions with natural or demographic challenges (Article 174 of the TFEU). Member States are required to coordinate their economic policies to achieve the objectives of cohesion. The Structural Funds, together with the European Investment Bank, play a crucial role in supporting cohesion (Article 175 of the TFEU).

Cohesion policy has been evolving down the decades in response to the changing socio-economic environment and EU's policy priorities and repetitive crises (Barca 2009; Bachtler and Mendez 2020; Maucorps et al. 2023; European Commission 2024a). For illustration, in the 2014–2020 period, more than 30% of EU funding was allocated to the 'social Europe' objective to support employment and inclusion measures, just over 25% was allocated to 'smarter Europe' to support investment in R&D, innovation, and competitiveness, and almost 20% was allocated to 'green Europe' and to 'connected Europe' (European Commission 2022). The priorities in 2021–2027 have an increased focus on a greener EU and on a just transition (European Commission 2024a,b).

For the next EU long-term budget, starting in 2028, the Commission plans to further prioritize funding and reform of cohesion policy to strengthen convergence, growth, and competitiveness. A key proposal is a budget further tailored to local needs, with National and Regional Partnership Plans that will guide investments and reforms, ensuring targeted impact where it matters most. The EU aims with this new reform to maximize the impact of every euro it spends, focusing on EU priorities and objectives where EU action is most needed.[3]

13.2.1 Principles

The EU Cohesion Policy is guided by key principles such as solidarity, concentration, programming, and partnership.

Concentration refers to the allocation of resources, which involves geographical concentration, with a focus on the poorest regions and countries, and thematic concentration, with a focus on a limited number of key objectives, including smart, green, and digital transformation.

Programming involves funding multiannual national and regional programmes aligned with EU objectives and priorities and negotiated with and approved by the European Commission. Funds must be spent by the end of the third year after their allocation (known as the $n + 3$ rule). For example, the funds committed for 2022 must be spent by the end of 2025.

Partnership is a collective process involving authorities at European, regional, and local levels; social partners; and civil society organizations, also referred in the literature as multilevel governance (Hooghe 1996; Bache 2008; Piattoni and Polverari 2016). It applies to all stages of the programming process, from design, through management and implementation to monitoring and evaluation through their participation in the monitoring committees. This partnership approach ensures that action is adapted to local and regional

3 On 16 July 2025, the Commission presented its proposal for post-2027 MFF amounting to almost EUR 2 trillion (or 1.26% of the EU's gross national income on average between 2028 and 2034): EU budget 2028–2034. The publication was preceded by public consultations and impact assessments and will be followed by inter-institutional negotiations leading to adoption by the end of 2027 at the latest.

needs and priorities and is guided by the European Code of Conduct on Partnership. Support to capacity building for partners is also ensured through the EU funds, with Member States required to allocate an appropriate amount of their resources to this purpose.

In the name of solidarity, the EU Cohesion Policy has, for example, responded with initiatives such as REACT-EU, the Recovery Assistance for Cohesion, and the Territories of Europe, aimed to address the economic and social impact of the pandemic on EU regions and to support their recovery and resilience.

13.2.2 Beneficiaries and Territorial Coverage

The EU Cohesion Policy benefits all Member States and regions, with over 70% of funding allocated to less developed regions. Country quotas ensure a fair distribution of funds, with each Member State receiving a specific allocation based on its economic and social needs. The national allocation is then divided among regions, based on categorizing regions in terms of their level of development, as indicated by their GDP per head. The policy also provides targeted support to urban areas, remote and island regions, mountainous areas, and outermost regions, recognizing their unique challenges and opportunities.

Moreover, while some funds benefit all EU Member States, namely the ERDF, the ESF+, and the JTF, the CF is allocated only for Member States with a Gross National Income per capita below 90% of the EU average and supports investments in areas such as transport and environmental priorities. In addition, the JTF alleviates the socio-economic costs triggered by the climate transition. It supports the economic diversification and reconversion of the territories concerned and helps people to adapt in a changing labour market.

13.2.3 Decision-Making, Programming, and Implementation

The cohesion policy framework is financially underpinned by the Multiannual Financial Framework (MFF), a seven-year budget plan that delineates the EU's spending priorities and allocations. The policy's scope is substantial, with over 392 operational programmes and 1.5 million projects supported during the 2014–2020 period, benefiting approximately 500,000 project beneficiaries (Bachtler 2022).

The policy decision-making, programming, and implementation of the policy involve several stages.

13.2.3.1 Decision-Making and Programming

The budget for the policy and the rules for its use are jointly decided by the European Council and the European Parliament based on a proposal from the Commission. The Common Provision Regulation (CPR) then provides a unified regulatory framework that applies to eight EU funds managed under shared arrangement, including cohesion policy funds.[4]

4 Regulation (EU) 2021/1060 of the European Parliament and of the Council of 24 June 2021 laying down common provisions on the ERDF, the European Social Fund Plus, CF, JTF and the European Maritime, Fisheries and Aquaculture Fund and financial rules for those and for the Asylum, Migration and Integration Fund, the Internal Security Fund and the Instrument for Financial Support for Border Management and Visa Policy.

Following the adoption of the MFF, the European Commission engages with Member States to formulate Partnership Agreements. These agreements specify the strategic goals and investment priorities that each country aims to achieve with its allocation of Cohesion Policy funding. Moreover, Member States are required to submit draft programmes that detail the planned interventions at national and regional levels. These may also include transnational cooperation programmes that involve multiple countries working collectively to tackle common challenges (e.g. Interreg programmes).

The Commission then negotiates with the national authorities to finalize the content of the Partnership Agreement and each programme. The programmes present the priorities of the country and/or regions, or the cooperation area concerned. Various stakeholders, including workers, employers, and civil society bodies, can participate in the programming and management of the programmes.

One essential component of the policy programming stage is ensuring that the right conditions for successful implementation are in place, representing a governance model that drives preparatory actions and ensures that necessary reforms and changes are made before investments start, enabling them to be carried out effectively. This is designed through enabling conditions, a key element of the 2021–2027 cohesion policy, which build on the ex ante conditionalities from 2014 to 2020 to ensure effective and efficient use of Funds (European Commission 2017). Enabling conditions must be respected throughout the programming period, creating conditions for success and allowing for reimbursement from the Union budget. They are linked to policy and strategic frameworks, as well as regulatory frameworks, to ensure high-quality strategic decisions and compliance with the EU acquis.

Another important component is the principle of additionality, a mandatory national or private co-financing which aims to guarantee a Member State's or beneficiary's commitment and ownership and ensures value for money in EU-supported interventions. The national co-financing rates of an instrument indicate the proportion of costs not covered by EU funding and they range from 15% to 60%, depending on the fund and the level of development of the supported region. However, there have been instances where Member States received 100% EU financing from cohesion policy funds because of exceptional circumstances, such as the financial crisis of 2008–2009, and for mitigating the effects of the COVID-19 pandemic and of the recent geopolitical crises.

13.2.3.2 Implementation: Programmes and Projects Selection and Delivery

The implementation of EU funding and related programmes is carried out through a system known as 'shared management', where the European Commission and the Member States collaborate to manage and implement EU funds. Under this system, both the Commission and the Member States share responsibility for the programming, monitoring, management, and control of Cohesion Policy funds. The Commission provides overall guidance and oversight, while the Member States are responsible for the day-to-day implementation and management of the funds at the national, regional, and local levels.

Hence, once the programmes are commonly agreed between the Member States and the Commission, Member States and their regions are responsible for implementing the planned actions. This involves selecting, monitoring, and evaluating hundreds of thousands of projects that meet specific strategic objectives and targets, as defined and adopted in the programmes. Programmes are supported by EU funds, currently the ERDF, CF, ESF+, and

JTF, aimed at addressing policy strategic and specific objectives. Projects are selected as to correspond and contribute to these main thematic objectives, each broken down into around 40 specific objectives.

The work of project selection and execution is organized by Managing Authorities in each country and/or region. All projects are selected by Managing Authorities (or on their behalf by, for instance, implementing bodies). Managing authorities establish and publish methodologies and criteria used for the selection of operations, approved by Monitoring Committees after having previously been reviewed by the Commission. In line with shared management, the Commission commits the necessary funds to allow Member States to start spending on their programmes (pre-financing). Furthermore, as the programmes are implemented by Member States authorities, the Commission monitors implementation, reimburses expenditure, and is ultimately accountable for the budget.

Both the Commission and the Member States are responsible for reporting and evaluating the programmes throughout and after the programming period.

Finally, a robust audit and control system is in place to ensure the legality and regularity of the expenditure. This system includes checks by national and EU audit bodies, safeguarding the financial integrity of the policy and maintaining public trust in its operations. The Member States have primary responsibility for setting up a management and control system, which complies with the common legal requirements, ensuring that this system functions effectively and preventing, detecting, and correcting irregularities. The Commission plays a supervisory role by satisfying itself that the arrangements governing the management and control system are compliant. It does so by verifying the effective functioning of this system and making financial corrections, where necessary.

In addition to the shared management system, the Commission implements specific initiatives under indirect management. The European Urban Initiative,[5] with a budget of EUR 450 million, supports cities of all sizes in application of sustainable urban development principles through building capacity and knowledge, supporting urban innovation by developing transferable and scalable solutions to urban challenges of EU relevance. Furthermore, to stimulate regional innovation projects in their commercialization and scale-up phases, the Interregional Innovation Investments (I3) Instrument, with over EUR 490 million available in the 2021–2027 programming period, supports cooperation between European regions, namely interregional innovation projects in their scale-up and commercialization phases, bringing together more and less developed regions.

13.3 Governance Arrangements of Nationally Financed Investments

Many Member States are either currently implementing or have recently implemented reforms of (parts of) their public investment management system. Many of these reforms tend to take an integrated approach to public investment management and address all

5 Article 12 of Regulation (EU) 2021/1058 of the European Parliament and of the Council of 24 June 2021 on the European Regional Development Fund and on the Cohesion Fund (hereafter Regulation (EU) 2021/1058.

stages of the investment cycle. Some Member States are using the Recovery and Resilience Facility[6] to accelerate reforms, while others may also rely on technical support from the European Commission via the Technical Support Instrument (TSI) managed by DG REFORM. Many use international expertise (e.g. IMF, OECD, and World Bank) for diagnosis and reform recommendations. Using interviews with reform managers, Belu Manescu et al. (xxxx) review these reforms and identify lessons learnt.

Against this background, the European Commission developed an analytical framework and started investigating the practices of managing investments in place in the Member States by means of an EU-wide survey (Belu Manescu 2021). In this framework, the investment cycle consists of five stages: strategic planning, appraisal and selection, budgeting, monitoring/implementation, and ex post review (including asset registers). The framework leverages previous work by the OECD, IMF, and the World Bank. It underscores the importance of feedback loops, especially in the first three stages, and the key role of the Ministry of Finance. The remainder of this chapter outlines evidence on these five stages, as discussed in more detail in Belu Manescu (2021, 2022, 2024).

Rules for managing public investment reported in the survey tend to vary with the project's size, sector, and source of financing. In general, more significant projects – traditionally in the transportation sector – are subject to stricter scrutiny. Similarly, EU-financed investments tend to follow stricter procedures than nationally financed ones. About half of the Member States reported significant differences in the management of EU-financed investments versus nationally financed investments. Key differences are reported across the full investment cycle, particularly with respect to rules for project appraisal and selection as well as for budgeting. Concerning budgeting, EU-financed investments, unlike nationally financed investments in most cases, are managed via a seven-year programming framework, which ensures that capital is available in a timelier manner throughout the project lifetime (see Section 13.2).

13.3.1 Strategic Planning/Initiation of Project

Strategic planning, of varying sectoral coverage, is in place in all Member States. A third of the Member States reported national strategies as their long-term strategic planning documents (Figure 13.1), covering all sectors and being accompanied by sectoral and sub-sectoral strategies. These countries are Bulgaria, Croatia, France, Greece, Ireland, Latvia, Lithuania, Poland, and Portugal. Half of the Member States predominantly referred to the transportation sector (occasionally defence as well), while four of them identified the EU Partnership Agreements as their only long-term planning document at the national level (Cyprus, Czechia, Hungary, and Slovenia).

Ireland offers one of the most transparent examples of an integrated spatial development and investment strategy in the EU. Effectively introduced in 2018, Ireland's strategic vision for the country, *Project Ireland 2040*, consists of two aligned and mutually reinforcing documents.

High-level visions guide public investment plans in Bulgaria, Croatia, Latvia, Lithuania, Poland, and Portugal. Like Ireland, the long-term development visions in

6 Regulation (EU) 2021/241 of the European Parliament and of the Council of 12 February 2021 establishing the Recovery and Resilience Facility.

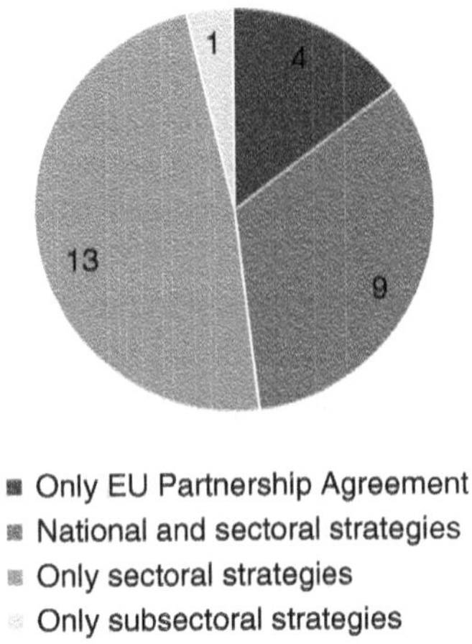

Figure 13.1 Strategic planning document in place in the Member States. *Source:* Belu Manescu (2024).

Latvia and Lithuania ('Latvia 2030' and 'Lithuania 2030') outline a 20-year development vision rooted in spatial planning and are accompanied by an investment plan with indicative expenditure allocations by investment programmes.

As the sector typically with the largest investment needs, transportation generally benefits from well-developed strategic planning and costed investment plans. Strategic plans in transportation often cover a horizon longer than 10 years. Planning for and delivery of transportation infrastructure often benefit from well-established specialized institutions. These include, for example, the National Transport Authority in Ireland, the National Transport Agency in Finland, and the Transportation Authority in Sweden. In Estonia, in 2019, the task of creating a unified administration for transport planning and investments and a unified administration for transport supervision was explored.

In several countries (e.g. Italy, the Netherlands, Ireland, and Latvia), operational planning involves a well-established process of consultation with all relevant stakeholders, including local and regional levels. In the Netherlands, the multi-year programme for infrastructure, public space and transport (MIRT) contains regulations on the cooperation in decision-making as well as requirements to receive funding for infrastructure or water projects.

13.3.2 Elements of Appraisal and Selection of Large Infrastructure Projects

Pre-appraisal is present in many Member States, yet with varying degree of complexity. Where it is present, it tends to be focused on preliminary costs and benefits, strategic relevance, and project rationale (in about 20 Member States). By contrast, the assessment of a full range of alternative solutions, potential for public–private partnerships and the planning of the appraisal process, are present in less than half of the Member States (Figure 13.2). Potential risks, and especially environmental sustainability risks, are considered at pre-appraisal in about 15 Member States.

Considering the alternative of not pursuing a project is a fundamental principle of pre-appraisal, yet only present in a few Member States. For example, the Swedish Transport Administration uses a four-step principle to investment planning, consisting

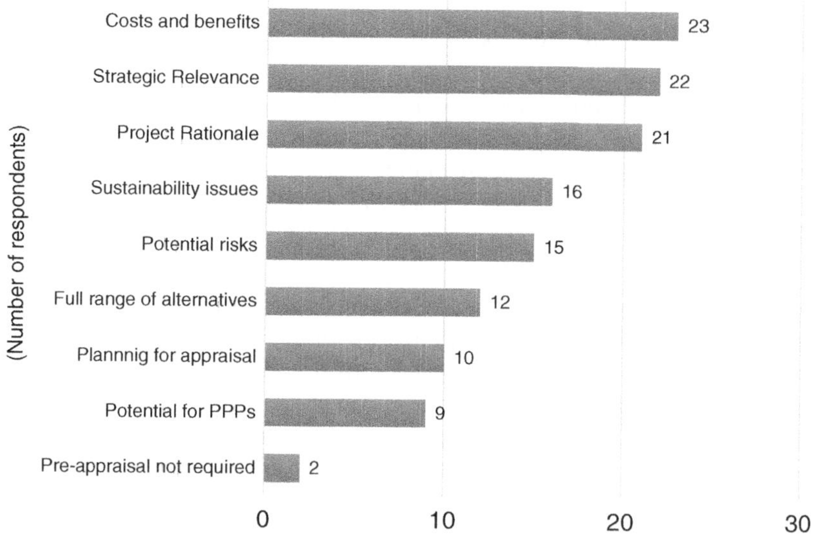

Figure 13.2 Pre-appraisal must include the following elements (survey). *Source:* Belu Manescu (2024). Note: PPP refers to public-private partnership.

of rethink, optimize, repurpose, and build new. Only after steps one to three have been considered can a new construction be built. A similar requirement is also found in the Netherlands and Ireland. In the EU, this requirement to consider and compare several alternatives to achieve the same objective is present in 12 Member States (Figure 13.2).

Importantly, pre-appraisal requires sufficient details on costs and clarity on the sources of financing. Pre-appraisal includes details on the life cycle costs of the project, i.e. management and maintenance costs. It is also when possibilities for pre-financing by local authorities and when private partners may be considered. In the Netherlands, 75% of the project funding must be available and committed before the pre-appraisal (called the 'exploratory phase') can start. In addition, co-funding by regional governments can be required in certain situations.

13.3.2.1 Appraisal

Appraisals provide detail and understanding of life cycle costs as well as outline the implementation plans. The implementation plans will elaborate on relevant details of costs, roles, and responsibilities as well as the implementation strategy, including interlinkages between programme components. In some countries, for example the Netherlands, 100% of financing must be available for projects to move to the appraisal phase.

According to the survey, project appraisal requirements tend to follow the principle of proportionality, with more stringent assessments for larger projects. A vast majority of Member States use a standard methodology for appraisal, at least in some sectors. Central support and development for appraisal/selection guidelines is available in about half of the Member States, especially in relation to EU-financed investments. Such bodies tend to be placed under the authority of the Ministry of Finance (e.g. Cyprus), Ministry of development (e.g. Greece and Czechia), and Ministry of Transport (e.g. Finland) or directly under the responsibility of the government (e.g. Lithuania). Furthermore, in some Member States, there are dedicated bodies in charge of carrying out appraisals for significant projects (e.g. Italy and Slovak Republic).

While appraisal refers to the methodologies and procedures to assess a project along all relevant dimensions (financial, social, environmental, sustainability), selection is concerned with the criteria and methods to compare findings across these dimensions, thereby reducing the number of eligible projects. In the EU, most often, comparisons across projects take place at the ministerial level rather than at the central level.

13.3.2.2 External Assurance Process

In the EU, independent external reviews are relatively common during appraisal, especially for transportation projects or for EU co-financed investments (Belu Manescu 2022). Only six Member States report to not have such processes in place. Such reviews are carried out by international experts (e.g. Portugal for research projects in tertiary education), consulting companies (e.g. Denmark for all infrastructure projects), independent agencies (e.g. Finland and Austria mostly for transportation projects), or by dedicated units under the Ministry of Finance (e.g. Cyprus, Romania, and Slovak Republic for all projects above a certain size). Ireland reports a specific system of independent expert reviews at two stages in the project life cycle (approval in principle and prior to tendering), supported by one network of experts and one advisory group. The network or panel of external assurers reviews the projects in all policy areas, while the Major Projects Advisory Group advises the Department of Public Expenditure, National Development Programme Delivery and Reform in reviewing the projects. Several Member States indicate that independent reviews are mandatory for certain EU-financed projects.

13.3.3 Budgeting Aspects

Multiannual budgets are indispensable tools for effective capital allocation. Multiannual investment budgets with a clear breakdown on maintenance needs, capital costs, and the available investment space provide clarity on the ongoing investment portfolio and the progress of projects each year. Committing (scarce) funds at the planning phase (before actual implementation) increases responsibility at the selection phase and avoids spending resources on unpromising projects.

13.3.3.1 Budgeting to Complete Projects in Construction

Multiannual budgets offer transparency on projects at different stages of development. Multiannual capital estimates at the programme/project level (based on implementation plans and updated every year) help keep track of capital needs over the medium term.

In the EU, forward-looking estimates of capital expenditure for the next three years are the most common information and are typically presented as an annex to the budget. 14 Member States report multiannual commitment appropriations (Figure 13.3), which are common in the transportation sector. Yet, some of these appropriations are only indicative, which can lead to large deviations between appropriations and execution. For example, in Germany, the Federal Ministry of Transport and Digital Infrastructure draws up a five-year financial planning document for the realization of the

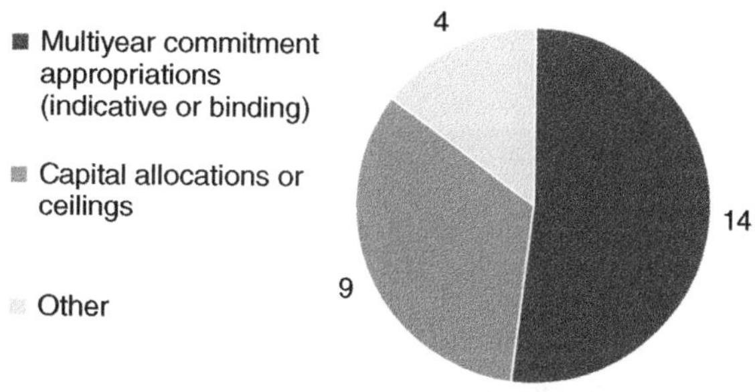

Figure 13.3 How is capital investment budgeted for? *Source:* Belu Manescu 2024.

Federal Transport Infrastructure Plan. The financial planning document contains detailed information about estimated costings for ongoing and future projects. Further, it stresses the capital needs to facilitate appropriate funding for maintenance, renewal of assets at the end of their life cycle, and creation of new assets.

Finally, EU-financed investments tend to enjoy a high level of protection over the years and take priority over other investments (see Section 13.2).

In very advanced systems, there are restrictions to ensure that approved projects do not exceed the available fiscal space. Such restrictions help lock-in funds to specific projects, thereby limiting the possibility to divert funds from existing projects to new projects. In very advanced systems, multiannual commitment appropriations, which are a formal set of legal restrictions and part of the legislative process for budgeting, help in this way Kim et al. (2020). This is in fact a dual system of cash and commitment appropriations, under which each line ministry receives two types of appropriations per year in the annual budget. The first is an appropriation for payments for the next budget year, and the second is an appropriation for commitments for each year for the entire duration of the projects. By controlling the full cost of ongoing projects over their lifetime, ministers can keep the stock of approved projects at an affordable level. Only few EU Member States have such a system in place (e.g. France, the Netherlands, Sweden, and Denmark), usually for larger projects in key sectors. Parliamentary approval is required for both the annual budget and multiannual commitments. For smaller projects, funding is provided as part of the annual budget bill, which the line ministries can prioritize without the need for further parliamentary approval.

Carry-over arrangements allow for unspent funds to be shifted to the following year. This is particularly useful in multiannual investment projects to protect slower-than-expected executed projects from losing funding (Belu Manescu 2022). While this practice tends to be restricted with current expenditure, as it carries some risks (Belu Manescu 2021), it is quite helpful in multiannual investment projects. For example, the Netherlands, Estonia, and Austria have no cap on the amount of carry-over in investment projects.

13.3.3.2 *Budgeting for Maintenance*

Only eight Member States report having in place a standard methodology for estimating maintenance costs in the budget (Figure 13.4), while maintenance costs are found in budgetary documentation in more than half the countries. Some countries also flag that unlike maintenance, costs due to major improvements or renewals are not reported in budget

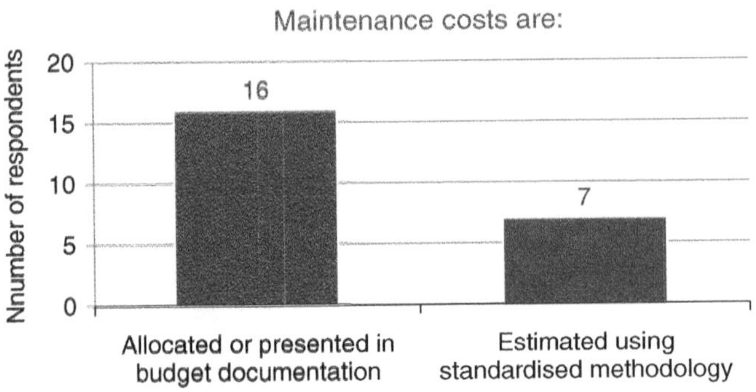

Figure 13.4 On maintenance costs. *Source:* Adapted from Belu Manescu (2022).

documentation for budget classification limitations (e.g. Poland). In the survey, it is reported that estimating maintenance costs for existing infrastructure is extremely difficult.

13.3.3.3 *Identifying New Fiscal Space*

The difference between the committed resources and the new resources available through the multiannual budget planning is the space for new investments. Capital expenditure ceilings are a critical instrument to provide clarity on new resources becoming available in the medium term for new projects, which can help prioritize new projects at the government/ministerial level. Such limits can thus play a crucial role in ensuring that the number of started projects stay within what is affordable, thereby avoiding the unwanted scenario of uncompleted projects because of the lack of funding. Such a system relies on accurate and timely monitoring of project implementation.

Capital allocations or ceilings are in place in less than half of the Member States (Figure 13.3). Such frameworks have different horizons and are binding at different degrees.

13.3.4 Monitoring and Implementation

All Member States report monitoring at the level of the implementing agency and line ministry, while additional monitoring is reported to take place also centrally in about half of the Member States. In Italy, the Ministry for Economy and Finance carries out real-time monitoring based on a joint project and programme management IT system connected to the contract system, which operates at the level of the line ministry. In countries with a large share of EU-funded investments, advanced electronic systems tend to be in place for EU-funded investments (e.g. Greece and Czechia).

For certain projects, usually more complex or sensitive projects, specific monitoring may be in place. For example, in France, major projects related to the national strategic investment plan (France 2030) are monitored by the 'Secrétariat Général pour l'Investissement', a body placed directly under the Prime Minister, while all sectoral projects are monitored by line ministries. Also in Latvia, the Council of Ministers may follow the implementation of certain significant projects, while in Poland, a dedicated 'Government Project Management Office' was established in the Chancellery of the Prime Minister's Office to monitor strategic projects. Finally, all EU-funded projects above a certain threshold are subject to specific monitoring (auditing) during the implementation process.

13.3.5 Ex Post Reviews

Three to five years after a project started delivering, a comprehensive ex post review can be carried out to assess strategic performance. As they require careful planning and are resource-intensive, systematic ex post reviews are rarely performed for all completed projects above a threshold. In almost all EU Member States, national audit offices carry out ex post reviews (also called performance evaluations) on a sampling basis. By contrast, in Ireland, the Public Spending Code requires all large capital projects and a proportion of the other capital projects to undergo ex post review, while in France, a similar requirement is in place for the investments in the France 2030 plan.

In few Member States, in addition to outcomes, the ex post reviews also assess the functioning of the public investment management system. They assess the extent to which the interaction between institutions, guidelines, and methodologies helped deliver intended results. Few Member States have such processes in place, such as Austria, Italy, or Spain (Belu Manescu 2022). Finally, for EU-funded investments, systematic requirements are in place for ex post reviews at priority axis level within each operational programme and have a feedback loop into planning.

A key final step, once a project is completed, is registering the asset in an official record of the assets owned by the state and its institutions. Some of the main functions of an asset register are to provide information on (1) the condition of the asset with a view to determine maintenance needs for existing assets and (2) inform the decision on new assets development by providing an overview of existing assets. Full asset registers are reported in key sectors (e.g. energy or transport) or government entities and are used for planning in Estonia, Finland, Greece, Sweden, Italy, Denmark, the Netherlands, and Germany (Belu Manescu 2022). Most often, such asset registers are not public. Other Member States are in the process of developing or extending existing ones (e.g. Cyprus and Greece).

13.4 Conclusions

This chapter examines practices for the management of public investment in place across the EU Member States, with special attention to the management of the EU Cohesion Policy. It provides a brief history of the principles of Cohesion Policy rooted in the Treaty of the European Union as well as details on the shared management between the European Commission and the Member States' national and regional authorities. It then provides an overview of practices covering the planning, appraisal and selection, budgeting, implementation, and ex post reviews in the Member States.

Overall, the evidence gathered reveals that public investment management practices at the central level tend to vary with the country's institutional set-up, project size, sector, and source of financing. In general, more significant projects – traditionally in the transportation sector – are subject to stricter scrutiny. Similarly, EU-financed investments tend to follow stricter rules throughout the project cycle than nationally financed ones, thanks to the policy rules jointly decided by the European Council and the European Parliament based on a proposal from the Commission.

These findings also highlight some gaps with respect to good practices across the Member States. For example, in the early stages of planning, many Member States do not consider alternatives to the main project, including the 'Do nothing' alternative. On the budgeting side, detailed evidence on the horizon for the multiannual commitment appropriations, how binding they are, and the use of carry over arrangements is missing in many countries. Finally, ex post reviews are often carried out selectively by Court of Auditors, while reviews of how the system works are rare and the use of asset register is limited.

References

Bache, I. (2008). *Europeanization and Multilevel Governance: Cohesion Policy in the European Union*. Rowman & Littlefield.

Bachtler, J. (2022). Cohesion policy: where has it come from? Where is it going? *ECA Journal* 1: 7–12.

Bachtler, J. and Mendez, C. (2020). Cohesion policy. In: *Policy-Making in the European Union*, 8e (ed. M.A.-R.H. Wallace): 232–253. Oxford University Press.

Barca, F. (2009). *An Agenda for a Reformed Cohesion Policy: A Place-based Approach to Meeting European Union Challenges and Expectations. Independent Report prepared at the request of Danuta Hübner, Commissioner for Regional Policy*. Brussels: DG REGIO.

Belu Manescu, C. (2021). *Public investment management in the EU: key features and practices*. ECFIN Discussion Paper Series nr. 154.

Belu Manescu, C. (2022). *New evidence on the quality of public investment management in the EU*. DG ECFIN Discussion Paper nr. 177.

Belu Manescu, C. (2024). *The planning of public investments in the EU member states: long-term strategy, selection and budgeting issues*. DG ECFIN Discussion Paper Series 213.

Belu Manescu, C., Cariou-Reynell, E., and Dimitrias, L. forthcoming. *Recent reforms of public investment management in the EU: overview and lessons learnt*. DG ECFIN Discussion Paper Series.

European Commission (2017). *The Value Added of Ex ante Conditionalities in the European Structural and Investment Funds*. Publications Office of the European Union.

European Commission (2022). *Eight Report on Economic, Social and Territorial Cohesion*. Publication Office of the European Union.

European Commission (2024a). *Forging a Sustainable Future Together – Cohesion for a Competitive and Inclusive Europe – Report of the High-Level Group on the Future of Cohesion Policy*. Publications Office of the European Union.

European Commission (2024b). *Ninth Report on Economic, Social and Territorial Cohesion*. Publication Office of the European Union.

Hooghe, L. (1996). *Cohesion Policy and European Integration: Building Multi-Level Governance*. Oxford University Press.

Kim, J.-H., Fallov, J.A., and S. Groom, (2020), "Public Investment Management Reference Guide", (Eds), The World Bank, Washington DC.

Maucorps, A., Römisch, R., Schwab, T., and Vujanović, N. (2023). The impact of the green and digital transition on regional cohesion in Europe. *Intereconomics* 58 (2): 102–110.

Piattoni, S. and Polverari, L. (2016). *Handbook on Cohesion Policy in the EU*. Elgar Publishing.

Ruiz Rivadeneira, A.M. and Mcmaster, P. (2023). *Mapping Institutional Arrangements for Infrastructure Governance in OECD Countries*. Paris: OECD Working Papers on Public Governance.

14

On the No-Objection at the World Bank: How a Practice Can Tell Us a Lot About Project Governance*

Lavagnon Ika[1] *and Javad Bakhshi*[2]

[1] *Telfer School of Management, University of Ottawa, Canada*
[2] *School of Business, University of Wollongong, Australia*

This chapter examines the practice of 'no-objection' at the World Bank – a formal mechanism by which the Bank's project supervisor signals to the recipient country government's project manager that the Bank does not oppose a proposed project transaction. While it may appear procedural, the no-objection practice speaks volume about project governance. It stems from the distinct roles of the World Bank, which provides funding and requires accountability, and the borrower country, which implements the project and maintains ownership. For project practitioners, this chapter offers key takeaways. Navigating the tensions between the funding agency accountability and the borrower country ownership requires a clear grasp of how oversight is exercised and what compliance entails. The no-objection is not just a checkpoint, but it is also a governance signal embedded in a multi-level system – project, organizational, and international – that shapes how international development goals are pursued and delivered. Understanding this dynamic is essential for both project supervisors and project managers who must look beyond the immediate project and consider broader governance layers to deliver successful projects. Indeed, securing a no-objection is not a rubber-stamp exercise; it requires alignment with procurement rules, transparency standards, and strategic objectives. Missteps can lead to delays, funding withdrawals, or reputational damage. The art lies in toeing the line for project managers and learning not to say 'No' for project supervisors.

14.1 Introduction

> 'The two shortest and oldest words are the ones that require a lot of thought: Yes and No'.
>
> – Pythagoras

* We are grateful to John Richard Heath, Adjunct faculty at American University and former World Bank staff member for his helpful comments on an earlier draft of this chapter.

Governance Frameworks for Major Public Projects: International Practices and Experiences, First Edition.
Edited by Vedran Zerjav, Morten Welde and Gro Holst Volden.

Consider, for example, the following comments regarding a contract award: 'We have no-objection...Please forward us a copy of the signed contract as soon as possible' (World Bank 2006). No-objection: That's the way the World Bank says 'Okay' or, more accurately, does not say 'No' to a recipient country to which it lends money for a project. Clearly, the no-objection is an institutionalized 'practice' in World Bank project oversight. Practices are often criticized but rarely interrogated. In this chapter, we ask: What does the no-objection practice entail? How does it work? And most importantly, what does it tell us about the role of governance in supporting the delivery of World Bank-funded projects?

We explore what the no-objection practice means in terms of governance in project settings. We reveal that a multilevel governance is at play in World Bank-funded projects, as it seeks to (1) provide checks and balances to steer project execution (project level); (2) enable projects to contribute to strategy implementation at the world's leading development agency (organizational level); and (3) foster collaboration with other development agencies, states, and non-governmental organizations (NGOs) (international level) (Müller 2016; Bakhshi 2023). We thus suggest that the no-objection practice stems from the separation of roles and responsibilities between the World Bank, who funds the project and requires accountability, and the borrower country, who implements it and maintains ownership. We conclude that if project supervisors and managers are to be fully successful in their collaboration and enhance the delivery of the benefits of the project and its sustainability, they should go beyond the project and consider other layers of governance like their World Bank counterparts.

14.2 The No-Objection and Its Governance Implications at the Project Level

International development projects rarely follow a simple principal-agent model; instead, they involve overlapping roles and responsibilities, creating both principal-principal and principal-agent challenges (Winters 2019). The World Bank funds projects, but it does not implement them (Ika and Feeny 2022). That task falls to national authorities, who operate with some degree of autonomy. The Bank's oversight is exercised through structured checkpoints, not direct control. The planning and implementation of World Bank projects occur through a complex network of stakeholders (Diallo and Thuillier 2004):

- the national project coordinator (the recipient country government agent tasked with the implementation of the project or hereafter the [recipient country government] project manager);
- the Task Team Leader (the World Bank staff member responsible for supervising the project or hereafter the World Bank project supervisor);
- the counterpart supervisor in the recipient country where the project is being implemented (typically a person deemed by the government to have the necessary skills, e.g. a high-ranking civil servant in some cases);
- a steering committee;
- suppliers of goods and services, including subcontractors, experts, and consultants;

- beneficiaries; and
- the population at large.

The 1000 km-long and four billion-dollar Chad–Cameroon pipeline project, where the World Bank sought to ensure that oil revenues served to actually reduce poverty in Chad, illustrates complex relationship dynamics across this network of stakeholders. Despite extensive oversight – including a watchdog committee, an international advisory group, and an external compliance monitor – governance failed to prevent misuse of oil revenues, which were diverted to military spending (Ika and Saint-Macary 2012). This underscores the limits of formal structures when national sovereignty and donor expectations collide.

This example highlights the importance of the separation of roles between the project manager and the World Bank supervisor, reflected in a principal-agent relationship, where the influence is exerted through compliance mechanisms rather than command. While the supervisor lacks formal authority over the manager, they are responsible for ensuring compliance with Bank guidelines (Müller 2016). This oversight is exercised through mechanisms like the no-objection notice, which allows the Bank to influence decisions without direct control (Diallo and Thuillier 2004; Ika and Saint-Macary 2012).

Although the project manager is not under the formal authority of the World Bank project supervisor, the latter must ensure that the World Bank's guidelines are strictly followed by the former. Hence, the World Bank project supervisor does not interfere with the day-to-day management of the project. Yet, they expect to be kept abreast on every key step of the project that is undertaken (Diallo and Thuillier 2004). But how does this separation of roles between World Bank project supervisors and project managers work in practice? How can a World Bank project supervisor oversee a project for which the project manager is not under their direct authority? In short, this is done through 'no-objection notices' (Ika and Saint-Macary 2012). So, what is a no-objection notice?

A no-objection notice is a formal confirmation from the World Bank project supervisor to the project manager, confirming that the Bank does not oppose a project transaction (World Bank 2006). The project manager, representing the client government, must request this clearance for key steps such as terms of reference, procurement, and contract awards (Diallo and Thuillier 2004). For example, when hiring a consultant, the supervisor reviews the documentation and, if it meets requirements, issues the no-objection (World Bank 2006).

While data on refusals is limited, the Bank may withhold a no-objection if procedures are not followed, planning is weak, or bidders are unqualified or ineligible (Ika and Saint-Macary 2012). If the government proceeds without approval, the Bank can declare misprocurement and cancel the related funding (World Bank 2006). To avoid this, the process must be repeated until compliance is achieved (Diallo and Thuillier 2004). For project managers, securing a no-objection requires strict adherence to Bank procedures, cultivating a culture of compliance. Meanwhile, Bank supervisors, lacking formal authority over project managers, refine their ability to influence outcomes without explicitly saying 'No' (Ika and Saint-Macary 2012).

More broadly, the no-objection mechanism reflects a multi-tiered governance model. At the project level, it ensures checks and balances. At the organizational level, it aligns execution with the Bank's development strategy. And at the international level, it facilitates

Table 14.1 Governance insights underpinning the no-objection practice.

Governance layer	No-objection explanations
Project governance	The no-objection is a project supervision practice and thus provides checks and balances to support the project at a distance. It stems from the separation of roles and responsibilities between the World Bank project supervisor and the project manager.
Organizational governance of projects	No-objection is an institutionalized practice that enables projects to contribute to strategy implementation from the funding organization's perspective. It stems from the supervisory role of the funder as the project is being implemented and helps ensure their accountability.
International governance of projects	The no-objection is an institutionalized practice that stems from the separation of roles between the funder of a project, which needs accountability, and the borrower country, which seeks ownership. It suggests a need for the world's leading development agency to collaborate with other development agencies, states, and NGOs to help support the delivery of a series of projects in recipient countries.

coordination among donors, governments, and NGOs (Müller 2016; Bakhshi 2023). For practitioners, this means that successful delivery depends not only on managing tasks but also on understanding the governance architecture in which those tasks unfold. Those levels are further elaborated in Table 14.1.

14.3 The No-Objection and Its Governance Implications at the Organizational Level

Governance at the project, programme, portfolio, and the board levels (Müller 2016) provides the World Bank with the means to enable strategy implementation. It sets the criteria for project selection, outlines the project's strategic objectives, defines the project's organizational structure, establishes how to align project decisions with objectives, and aims to ensure that projected benefits are realized (Bakhshi 2023). Such goals and objectives can be multifaceted and complex, for example eliminating extreme poverty and boosting shared prosperity and, more recently, a commitment to addressing issues related to global public goods such as climate change and pandemics (Ika et al. 2020).

To achieve such strategic goals in collaboration with recipient country governments, the World Bank uses projects as key instruments for policy implementation. Again, the World Bank funds but does not implement projects itself (Baum 1978). Still, over the decades, as we show later, there has been a steady accretion of hoops that World Bank project supervisors must jump through when preparing and overseeing projects. The World Bank is actively involved in the strategic planning phase, shaping the business case for a project, supporting the identification of the project, and advising government analysts on its definition, mission, and vision, as aligned with the recipient country's national development plan or partnership framework. The World Bank also has a say when it comes to assessing the economic, financial, technical, institutional, environmental, and gender aspects of the project (World Bank 2024a).

From this conceptual definition, the project goes through a standard design phase where needs, options, feasibility, and risks are analysed, and ultimately reaches an operational planning phase where estimates of duration and cost as well as monitoring measures are created. Thus, a strong front end characterizes World Bank-funded projects (Picciotto 2020). A decade may elapse between the identification and the subsequent approval of the project and its actual implementation start-up in the recipient country. For example, it took seven years between the Chad–Cameroon pipeline project's identification and its formal approval by the World Bank (cf. 1993–2000) (Ika and Saint-Macary 2012). Prior to the implementation of start-up, the World Bank project supervisor prepares a supervision plan that includes the arrangements, with the recipient country staff, in terms of monitoring, reporting, and evaluation. Then, the project enters the implementation phase where it is subject to strong and bureaucratic guidelines from the World Bank, as mentioned earlier. During the implementation phase, the World Bank project supervisor, who often supervises a portfolio of four or five projects at the same time, generally in the same field, but not necessarily in the same country, mainly oversees and monitors the project implementation from the World Bank headquarters at Washington, DC, or from the World Bank country office and through field visits to the recipient country (Ika 2015).

Hence, as mandated in its guidelines, the World Bank has devised a bespoke project life cycle, an organizational governance framework to design, prepare, implement, and supervise projects with clear-cut roles and responsibilities between the project supervisor and the project manager (see Figure 14.1). This framework, which has evolved since the 1970s (Baum 1978), proposes an orderly progression through six stages from identification to preparation, appraisal, negotiation/board approval, implementation, and completion/validation and evaluation (see World Bank 2024a).

In the identification phase, the government analysts prepare their own country partnership frameworks with the assistance of the World Bank and, in so doing, identify their priorities in terms of the three main goals of the World Bank, i.e. eliminating extreme poverty, boosting shared prosperity, and addressing global challenges. Both the World

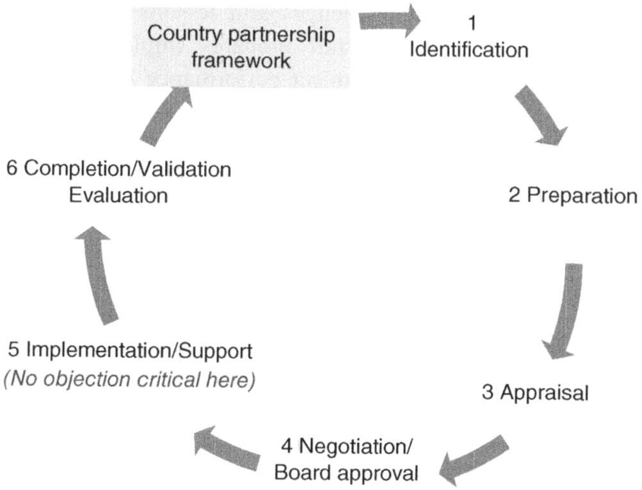

Figure 14.1 World Bank project cycle. *Source:* Adapted from World Bank (2024a).

Bank and the recipient country then agree on an initial project concept and beneficiaries, the preliminary feasibility and risk assessment of the project including environmental and social concerns, and a suitable timetable for the project preparation process. In the preparation phase led by the government project supervisors, assisted whenever necessary by consultants and private sector organizations, the World Bank task team offers advisory and technical support in terms of project design, feasibility, risk, and assessment of the capacity of the implementing agencies and reaches an agreement with the recipient government on both the content and design of the project and project management arrangements including reporting, monitoring, and evaluation. The appraisal stage is where both the World Bank task team and the government project supervisors thoroughly review the project design and reach an agreement on the viability of all the aspects of the project and identify any unfinished business required for final Bank approval.

Once approved by the World Bank's Board of Executive Directors at the end of the negotiation and approval stage, the project enters its implementation phase, undertaken by the government or more specifically their appointed project manager. The implementation phase is broken down into different stages as determined by the preparation phase. The project is financed in chunks provided by the World Bank and/or other funders if necessary. Each chunk is released, aborted, or scaled down by World Bank management depending on the conclusions of the supervision report designed to inform the management review. At this stage, the World Bank task team offers monitoring services to help reduce the odds of delays, cost overruns, coordination breakdowns, benefit shortfalls, and beneficiary dismay. The World Bank task team thus provides implementation support to improve results, help manage risks, and increase institutional development in the recipient country. As the World Bank suggests, a review of the project is held at least twice a year: 'Changing circumstances, project delays and unexpected events can sometimes require adjustments to the project design, such as implementation arrangements or even objectives, resulting in a restructuring' (World Bank 2024a).

Once completed and closed, the project moves to the completion/validation and evaluation stage where the project's outcomes are assessed in compliance with the World Bank's policies and guidelines, along with challenges and lessons learned to sustain benefits realization. While the World Bank staff prepares Implementation Completion and Result Reports, which include initial project performance ratings, the Independent Evaluation Group (IEG) staff validates all these self-evaluations and hence evaluates World Bank-funded projects, programmes and activities by examining their achievements against stated objectives (e.g. Ika and Feeny 2022). Thus, a strong back-end characterizes World Bank-funded projects (Picciotto 2020). Figure 14.2 summarizes the above discussion and depicts what could be considered the World Bank stage-gate process and key milestones.

Because project supervision at the World Bank may be understood as project sponsorship or 'the supervision of project management' (Ika 2015), the arms-length governance principle implied in the no-objection notice can help explain the apparent disconnect between planning and implementation that seemingly prevails at the World Bank.

As taking up the implementation of projects in addition to their planning would both jeopardize ownership of the projects by the sovereign governments and add administrative costs for the World Bank, the World Bank is stuck with the practice of no-objection

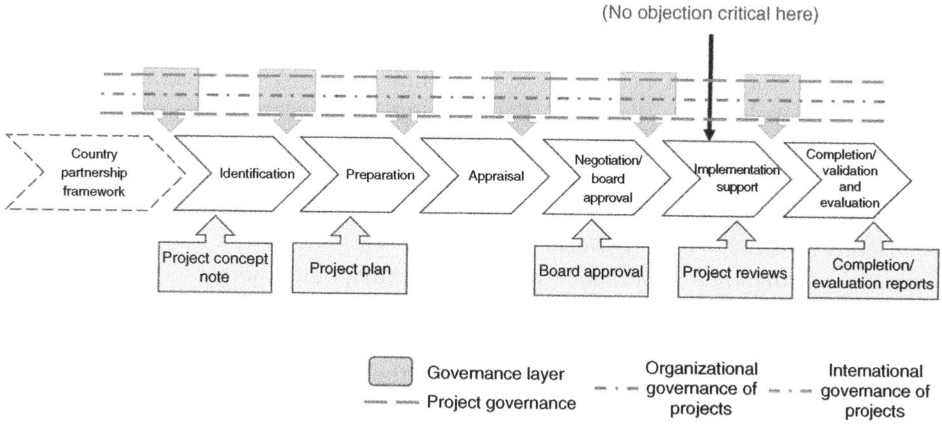

Figure 14.2 World Bank stage-gate style process and key milestones.

for the foreseeable future. Otherwise, its supervisory role and its accountability function will not be fulfilled (Ika and Saint-Macary 2012). This is why the monitoring and evaluation function is critical and Results Based Management (RBM) – a forerunner of benefits management frameworks, which looks beyond activities and outputs and focuses on benefits and outcomes – remains the dominant approach for managing World Bank-funded projects, despite its well-documented rigidities and dysfunctions (Ika et al. 2020).

14.4 No-Objection and Its Governance Implications at the International Governance Level

As we have shown, the use of the no-objection notice stems from the separation of roles and responsibilities between the World Bank project supervisor and the recipient country government project manager and the need to ensure accountability for the World Bank. In this section, we offer an explanation, not just from an interorganizational perspective but also from an international network governance perspective by focusing on the key issue of ownership of the project for the recipient government (Lie 2019), which we discuss later. Put differently, we refer to the governance of interorganizational networks that are formed by three or more partner organizations (Manning 2017) in two or more countries to increase their joint performance in the delivery of projects (Ika et al. 2020). Development agencies, such as the World Bank, repeatedly work together and draw from local and international networks of partners all over the world depending on where the projects are implemented (Manning 2017). For example, the Chad–Cameroon pipeline was an international collaboration between the World Bank, the governments of Chad and Cameroon, and the oil consortium including the U.S.-headquartered ExxonMobil and Chevron and Malaysia's Petronas (Ika and Saint-Macary 2012). Other stakeholder groups included NGOs in the United States such as the California-based Rainforest Action Group and other local NGOs such as the Cameroon-based Center for Environment and Development.

At the time when key observers argue that the World Bank needs a significant overhaul to tackle some of the world's most pressing issues such as global poverty reduction, climate change, and global pandemics (Ravallion 2016; Harvey and Greenfield 2022), which require not just a country or region focus but also global governance (Ika et al. 2020), if you ask what the World Bank does – a question raised by the Associate Director, Projects of the World Bank, Warren C. Baum, about half a century ago – you are then and now likely to hear that the World Bank funds and supervises projects in recipient countries (Baum 1978; Ika and Feeny 2022). Since 1947, the World Bank has funded over 12,000 projects (e.g. Ika and Feeny 2022). In 2023, the World Bank had, under supervision, 1778 projects worth $324 billion (World Bank 2024b).

From an international governance of projects standpoint, there prevails a separation of roles between the World Bank and the government, hereafter, the borrower. While supervision is the responsibility of the World Bank, planning and implementation fall to the borrower (World Bank 2024a). This begs the question: Why does the borrower need a no-objection from the World Bank? The World Bank, as a multilateral agency in the development sector, raises billions of dollars per year from donor nations such as its largest shareholders, the United States, France, Great Britain, and Japan, and then doles out loans to recipient countries for funding projects. On average, the World Bank lends $67 billion and approves 400 projects every year (World Bank 2024b). These projects can be infrastructure projects such as bridges or agricultural projects such as fish farms, or 'process' projects including education, health, policy advice, and capacity building projects or even emergency projects such as post-disaster or post-conflict reconstruction projects (Ika and Feeny 2022). While projects may typically last three to five years, the funding can be up to 10 years. But, for what purpose?

Development agencies such as the World Bank aim to enact social change in the life of poor people and thus contribute to economic growth, poverty reduction, or grand challenges curbing notably in recipient countries. World Bank-funded projects are, by their very nature, social, and political undertakings; social in the sense that the aim is to directly or indirectly improve the wealth of the population, and political because leaders may push for pet projects in favour of specific locations and target groups, for career and political gains. Yet, a key part of the World Bank's mandate remains the financing and supervising of projects (Baum 1978). While the scope of the World Bank's business has changed over time, it still fundamentally lends money to recipient countries to deliver projects (Ravallion 2016). And in return, the borrower needs a no-objection to proceed with the use of the loan during project implementation (World Bank 2006).

But, in contrary to many projects for which the clients pay for the project and benefit from them, the World Bank pays for but does not receive project deliverables. The borrower country government usually does so on behalf of the target beneficiaries of the project. The World Bank does not want the funded project to be considered its own. Otherwise, this may pose an ownership problem and the project risks not reaching its economic, social, or environmental objectives. The Paris Declaration, for example, stipulates the ownership agenda, which expects partner countries to exercise effective leadership over their own development policies, strategies, and actions (Paris Declaration 2005). The ownership agenda is supported by the participatory, partnership-based, and capacity-building approaches that are ever-present in practice (Picciotto 2020). So, a World Bank-funded project is not a World Bank project. However, the borrower and its project implementation unit must conform to stringent bureaucratic and transactional World Bank's guidelines for financial reporting, environmental assessment, procurement procedures, and monitoring (Ika 2015).

The purpose is to ensure that rigour and transparency are maintained in the award of contracts and performance of tasks and that the project is implemented as planned or appropriately modified in light of its changing environment, so that its economic, social, or environmental objectives are achieved (Picciotto 2020). And this is a governance challenge. One of the ways the World Bank does this is through no-objection. Interestingly, if the no-objection is a powerful tool in the hands of the World Bank, it is perhaps because 'As one of the world's richest institutions, the World Bank has much more power and influence than the poor countries and institutions that implement the project, and even more than the beneficiaries who receive its deliverables' (Ika and Saint-Macary 2012, p. 431).

As Lie (2019) suggests, the ownership agenda is supposed to get rid of the asymmetrical relationships that percolate through the governance and delivery of projects. However, upward accountability often prevails at the expense of downward accountability in practice. As Bandé et al. (2024, p. 1) note, using the Chad–Cameroon pipeline as an example, the project 'met its deadline, amidst social and environmental challenges; however, it was seen as a success for the oil consortium, the Chadian government, the funders, suppliers and contractors, but not for the poor beneficiaries'. The tensions between the imperatives of accountability for the World Bank and ownership for the borrower may also lead to what we term 'accountability-for-results trap', a focus on tangible results by the World Bank, even if they are meaningless or dysfunctional.

More importantly, Lie (2019) offers a revealing case study of the World Bank–Uganda partnership and shows how the World Bank uses the ownership concept to make Uganda responsible for the World Bank policy priorities, also called 'prior actions', and to foster the use of technocratic governance mechanisms to govern at a distance. In this instance, much to the dismay of bilateral donors – also members of the World Bank and thus represented on its board – the main opposition candidate in Uganda was imprisoned. Amidst fears that their board of directors would not approve the US$150 million in budget support, the World Bank country and mission teams then unexpectedly cut their support to the country by one-tenth, officially invoking economic mismanagement and, in particular, 'concerns about expenditure overruns in the public administration budget'. This budget cut forced the Ugandan government to downsize the planned activities in their national development plan, which was previously agreed to by the World Bank (Lie 2019, p. 1120).

As the World Bank–Uganda partnership suggests, the tensions between accountability for the World Bank and ownership for the recipient country government are frequent. Therefore, the no-objection principle is also a policy tool to ensure the sovereignty of the recipient governments' decisions to implement the projects that best contribute to their own national development goals, rather than simply implementing a policy agenda imposed by the donor countries.

14.5 Conclusions

The chapter focuses on the scope and ramifications of the no-objection notice, which serves as a governance tool to ensure compliance with World Bank standards and processes, while also offering some ownership to the recipient country. It argues that if the World Bank institutionalizes the practice of no-objection, it is mainly because of the separation of roles and responsibilities between the World Bank, who funds the project, and the recipient (borrower) country, who implements it. This translates into the separation of roles and responsibilities between the World Bank project supervisor and the recipient country government project manager. The no-objection appears to be a project supervision practice, which stems from the need to ensure accountability of the project for the World Bank and ownership of the project for the recipient country government.

The chapter thus demonstrates that a triple process of governing projects, operating at the project, organizational, and international levels of governance, is at work and underlies the specific practice of no-objection at the World Bank. These layers of governance, respectively, provide checks and balances to steer project execution, enable projects to contribute to strategy implementation, and help the world's leading development agency to collaborate with other development agencies, states, and NGOs. Consequently, the chapter advances that it is only by looking at the interconnections among these layers of governance that supervisors of World Bank-funded projects can effectively support project execution and help project managers deliver project success and in particular sustainable benefits for the recipient country, the World Bank, other stakeholders, and beneficiaries. As navigating multilevel governance is critical for project success, project supervisors and managers can learn from their World Bank counterparts. So, let us welcome a no-objection to going beyond 'best practices' and, thus, better understand multilevel project governance for more success!

References

Bakhshi, J. (2023). The influence of governance and complexity on project outcomes in international development. Doctoral dissertation. The University Sydney School of Project Management, Australia.

Bandé, A., Ika, L.A., and Ouédraogo, S. (2024). Beneficiary participation is an imperative, not an option, but does it really work in international development projects? *International Journal of Project Management* 42 (1): 102561.

Baum, W.C. (1978). The World Bank project cycle. *Finance and Development* 15 (4): 10–17.

Diallo, A. and Thuillier, D. (2004). The success dimensions of international development projects: the perceptions of African project coordinators. *International Journal of Project Management* 22 (1): 19–31.

Harvey, F. and Greenfield, P. (2022). Climate-focused reform of World Bank could be done in a year, says Al Gore. *The Guardian*. https://www.theguardian.com/business/2022/nov/14/climate-focused-reform-world-bank-could-be-done-year-al-gore (accessed 17 December 2025).

Ika, L.A. (2015). Opening the black box of project management: does World Bank project supervision influence project impact? *International Journal of Project Management* 33 (5): 1111–1123.

Ika, L. and Feeny, S. (2022). Optimism bias and World Bank project performance. *The Journal of Development Studies* 58 (12): 2604–2623.

Ika, L.A. and Saint-Macary, J. (2012). The project planning myth in international development. *International Journal of Managing Projects in Business* 5 (3): 420–439.

Ika, L.A., Söderlund, J., Munro, L.T., and Landoni, P. (2020). Cross-learning between project management and international development: analysis and research agenda. *International Journal of Project Management* 38 (8): 548–558.

Lie, J.H.S. (2019). Local ownership as global governance. *The European Journal of Development Research* 31 (4): 1107–1125.

Manning, S. (2017). The rise of project network organizations: building core teams and flexible partner pools for interorganizational projects. *Research Policy* 46 (8): 1399–1415.

Müller, R. (2016). *Governance and Governmentality for Projects: Enablers, Practices, and Consequences*. Taylor and Francis.

Paris Declaration (2005). Paris declaration on aid effectiveness. Ownership, harmonization, alignment, results and mutual accountability. Paris, High Level Forum, February 28–March 2, 2005.

Picciotto, R. (2020). Towards a 'New Project Management' movement? An international development perspective. *International Journal of Project Management* 38 (8): 474–485.

Ravallion, M. (2016). The World Bank: why it is still needed and why it still disappoints. *Journal of Economic Perspectives* 30 (1): 77–94.

Winters, M.S. (2019). Too many cooks in the kitchen? The division of financing in World Bank projects and project performance. *Politics and Governance* 7 (2): 117–126.

World Bank (2006). *Consulting Services Manual. A Comprehensive Guide to the Selection of Consultants*. Washington, DC: The World Bank.

World Bank (2024a). *World Bank Project Cycle*. World Bank Project Cycle.

World Bank (2024b). FY25 World Bank budget text for public disclosure. https://documents1.worldbank.org/curated/en/099080824144031957/pdf/BOSIB196bde0ea0fb1b88215cf5c0e36a36.pdf.

Part III

Comparative Analysis and Lessons for Policy Implementation

15

The Country Frameworks Summarized

Gro Holst Volden, Morten Welde and Vedran Zerjav

Concept Research Programme, Norwegian University of Science and Technology (NTNU), Trondheim, Norway

This chapter provides a comparative overview of governance frameworks for major public sector investments across a selection of developed countries, the European Union (EU), and the World Bank. It examines how these frameworks have evolved to address challenges such as cost overruns, benefit shortfalls, and political interference, with the goal of enhancing project planning, accountability, and strategic success. By examining the scope, structure, and implementation of governance models in countries with diverse administrative traditions and institutional capacities, this chapter highlights both commonalities and contextual differences. It also reflects on the effectiveness of these schemes, their limitations, and the opportunities for future improvement in public investment governance.

15.1 The Countries Studied

The book examines governance models for major public sector investments and how these are currently organized and implemented in nine countries, the European Union (EU), and the World Bank. The nine countries are shown in Table 15.1. They are all developed countries with mature democracies and high-perceived standards of public sector governance. They all carry out numerous large and often complex investments annually, which are primarily funded by taxes. In Canada, we include a separate description of the governance scheme in the province of Quebec as an example of a well-developed province-level scheme.

Except for Quebec, we focus on governance arrangements introduced at the topmost level of government of the country. In Australia, Canada, and the United States, this refers to the federal level, whereas in unitary state countries, it pertains to the central government level. The topmost level is typically responsible for national infrastructure such as defence, major highways, railways, airports, national healthcare systems, and major ICT infrastructure (in countries with a federal system, the states are responsible for a large share of infrastructure delivery, while federal governments retain control for nationally significant infrastructure, including defence). All of the countries have local administrative levels

Governance Frameworks for Major Public Projects: International Practices and Experiences, First Edition.
Edited by Vedran Zerjav, Morten Welde and Gro Holst Volden.

Table 15.1 The included countries (population and GDP per capita).

Country	Governance level studied	Population (2023)	GDP per capita[a] (2023)
Norway	Country	5.6 million	US$101,031
The United Kingdom	Country	69.2 million	US$60,620
The Netherlands	Country	18.0 million	US$84,222
Ireland	Country	5.4 million	US$130,797
Denmark	Country	6.0 million	US$79,514
Sweden	Country	10.6 million	US$71,031
The United States	Federation	340.1 million	US$85,810
Canada	Federation	41.3 million	US$65,463
Australia	Federation and states	27.2 million	US$71,410

a) PPP method
Source: Data from the World Bank and Institut de la statistique du Quebec.

as well. In Ireland, the central government's scheme applies to local authorities' investment projects. In the other countries, the scheme may apply to local investments only to the extent that they receive government funds for specific projects or programmes.

The countries differ in size, population density, and natural conditions such as topography and climate. This implies that it may be more challenging for infrastructure projects to achieve a positive benefit–cost ratio in some countries than others. For example, Norway has a small and scattered population and challenging topography, but there is a broad consensus that people should be able to live in such locations as well.

There may also be national differences concerning trust in authorities and work culture, which may affect the governance arrangements. For example, in the Anglo-American tradition, there is a strong focus on personal responsibility and meritocracy, where individuals are rewarded based on their performance and achievements. By contrast, Scandinavian countries are characterized by a tradition of high social trust, openness, and flatter organizational structures and less focus on personal responsibility in government agencies. In Scandinavia, there is also a tendency to emphasize features of the system as opposed to the individual – particularly when things go wrong.

15.2 Establishment and Purpose of the Frameworks

While the World Bank has had governance arrangements related to its project-based loans and grants since its establishment in 1944, most of the countries only introduced their project governance models around 2000 and thereafter. The models were not introduced in a vacuum – most countries already had general budgeting and fiscal rules, and some have had general procurement regulations since the 1960s. Tools and methods such as cost estimation and cost–benefit analysis (CBA) were already in use, at least in some sectors. However, it was only in the 2000s that stage-gate models were introduced – often as loose models, which were later formalized with clearer requirements concerning appraisal and planning, quality assurance (QA), and approval at each stage.

Table 15.2 Main purpose of the project models.

Country	When established	Purpose of the first establishment of model	Later developments
Norway	2000	Improve tactical project success	Extended to focus on strategic project success as well
The United Kingdom	2000	Improve project success and accountability	Extended to focus more on strategic project success. Several reorganizations of the central unit responsible for the scheme
The Netherlands	2000	Ensure broad participation, commitment, and legitimacy of decisions. Ensure the use of CBA	More competition between transport modes introduced
Ireland	2021	Focus on economic appraisal, later also on risk management, assurance, and streamlined approval processes	
Denmark	Transport 2007 IT 2011	Improved cost control (referred to as a 'budget scheme')	The scheme for transport projects has widened over the years to include building construction projects. Model for IT projects introduced later
Sweden	No formal scheme	Need to streamline processes, and need to improve cooperation between local, regional, and state level	Increased focus on cost control in the earliest project phase. Regulations for IT projects introduced later
The United States	Varies (separate schemes in the c. 25 federal agencies)	The need to ensure proper management of cost, time, and risk (tactical and strategic project success)	All of the schemes have evolved over time
Canada	1978 for the 'Major Crown Projects', first full version in 1994	Focus on efficiency/economy	Several changes over time, such as more focus on value for money and accountability
Quebec (Canada)	2008	Improved tactical project success	The focus on societal needs and benefits has been strengthened over time
Australia	2006–2007	The need to govern large and risky projects and avoid politicization of projects	

The background to many of the schemes was negative experiences from past projects, most notably cost overruns and delays, as well as benefit shortfalls. When Norway, Australia, the Netherlands, Denmark, and Quebec introduced their schemes in the early 2000s, a key purpose was to improve the quality of appraisals and planning documents, streamline the processes, and improve tactical project efficiency. The Norwegian scheme was later expanded to encompass the more important choice of concept and focus on value for money. Other schemes, too, were broadened in similar ways. In some schemes, strategic project success was crucial from the start. For example, there has been a strong focus on CBA in the Netherlands, the United Kingdom, and Sweden. As discussed in the Dutch chapter, there was a big discussion in the 2000s in the Netherlands about the lack of, and often poor quality of, CBAs. With the governance scheme, CBA became compulsory in all decision-making processes for large infrastructure projects.

Only two countries, Sweden and the Netherlands, have expressed that the formal involvement of stakeholders at the local administrative level was also a motivation for their schemes. The Swedish and Dutch models clearly define involvement processes, in which different parties are encouraged to find solutions together. In both cases, this also ensures that those who express their needs and requirements take some responsibility, for example, through co-funding from local governments. In the case of the Netherlands, working with stakeholders is a cultural legacy of cooperation needed amongst polder communities to maintain the land reclaimed from the sea habitable.

Accountability is a key consideration in all spending, although this is not always explicitly stated as a motivation behind the schemes. Some schemes involve individuals taking clear ownership and responsibility for decisions made. In the United Kingdom, a Senior Responsible Owner (SRO) is a key leadership role in the framework, typically taken on by a senior executive within the ministry. The SRO owns the business case, makes major decisions, and ensures that risks are managed, and it is ultimately accountable for the project meeting its objectives. Similar roles are found in the schemes in Australia and Canada. However, this is not always accompanied by full transparency of project documents or ex post reviews.

The main purpose of the project models, when they were introduced and today, is summarized in Table 15.2.

15.3 Scope of the Frameworks

Most studied countries have a universal project governance model that applies to various sectors and project types. However, IT projects are sometimes treated differently by having a different model or lower thresholds. Furthermore, several models have exemption provisions, excluding state-owned enterprises, quasi-autonomous institutions, and local projects with central government funding. Only some models explicitly cover defence acquisitions, and only the UK model applies to the so-called transformation and service projects. The United States stands out, with no overriding scheme beyond general budgeting regulations and acquisition rules. Instead, each of the 25 federal agencies has its own model.

The schemes target the largest, most complex, and innovative projects. Most countries have threshold values that define 'large' projects, whereas 'complex' and 'innovative'

Table 15.3 Content and scope of the studied project models.

Country	Number of schemes	Sectors covered	Projects covered	Main components	Dissemination	Competence development as part of the model?
Norway	1	All except health and energy[a)]	Large[b)]	Stage-gate model, external QA	QA reports are published	Trailing research. Some training activities
The United Kingdom	1	All	Large, complex, or strategically significant	Stage-gate model, QA, formal roles, portfolio management	Delivery confidence ratings published	Education programme and accreditation scheme
The Netherlands	1	Infrastructure and water	Large[c)]	Stage-gate model, stakeholder representation (portfolio management)	Project documents published	Some training activities
Ireland	1	All	All, but degree of scrutiny depends on size[d)]	Stage-gate model, external QA, formal roles, (portfolio management)	Review notes are generally published, with recommendation that project documents are also published	Some training activities
Denmark	2	Transport/ buildings and IT	All/large[e)]	Stage-gate model, external QA	Project documents published	Some training activities for IT
Sweden	(2)	Transport and IT	All in principle	Stage-gate model, stakeholder representation	Project documents published	No
The United States[f)]	25	All	Most schemes have thresholds	Stage-gate model, QA, formal roles	QA reports are generally confidential	Some training activities

(Continued)

Table 15.3 (Continued)

Country	Number of schemes	Sectors covered	Projects covered	Main components	Dissemination	Competence development as part of the model?
Canada[f)]	1	All	All, but degree of scrutiny depends on size	Stage-gate model, mostly internal QA, formal roles	QA reports are generally confidential	Some training activities
Quebec (Canada)	2	All (IT is separate)	Large[g)]	Stage-gate model, QA		Some training activities
Australia[f)]	1	All except defence, which has a separate model (not covered in this book)	All, but the degree of scrutiny depends on size	Stage-gate model, internal/external QA, formal roles		

a) Includes all sectors in principle, but state-owned enterprises are exempted, which results in some sectors not being covered by the model.
b) All projects except IT: >1 billion NOK, IT: >300 million NOK.
c) €>112.5/225 million depending on municipality.
d) €>200 million.
e) Transport: all, but external QA only for >350 million DKK, IT: >15 million DKK.
f) Only schemes on the federal level are included here. States will normally have their own schemes.
g) >100/50 million CAD for road/others.

would normally require a more subjective assessment. The models often have general requirements for all projects, independent of size, whereas special rules and a higher degree of scrutiny apply to the largest projects.

The primary component is a stage-gate model that defines crucial decision points and requires appraisal and planning documents to undergo internal and/or external QA. Most models have one to three approval gates in the front-end phase. Management and governance roles, such as SRO, sponsor, and steering group, are described in some models. All the schemes apply to individual projects (and programmes), whereas the portfolio level is in most cases not explicitly covered.

This book focuses on governance models at the topmost level of government. Some (for example, Norway) are simple and focus only on the most crucial control points, while leaving the detailed requirements to line ministries and agencies that have their own models. Other government models are more detailed and focus on all the steps, with specific requirements concerning roles and responsibilities, tools and methods, templates, guides, etc. The latter strategy will ensure consistent project practices across ministries and agencies, but perhaps at the expense of local independence and flexibility.

Transparency on project appraisals, recommendations, and reviews can be difficult to obtain and is not necessarily part of the schemes. Some chapters in Part II of this book describe how the lack of transparency and good data can hinder accountability and legitimacy.

Capability development is rarely included, at least not in the form of formal training or project management education as part of the schemes. Rather, it is assumed that educational establishments such as universities provide the necessary training and that ministries and agencies hire skilled staff. However, in Norway, the Ministry of Finance funds a research programme to collect data and experiences and disseminate best practices. In the United Kingdom, the Infrastructure and Projects Authority (IPA) (reorganized from 1 April 2025 as National Infrastructure and Service Transformation Authority [NISTA]) supports public agencies in developing project and programme management capacity. The United Kingdom is the only country with education programmes to ensure high skills in government-funded projects (Table 15.3).

15.4 Parties and Roles

15.4.1 Main Parties and Roles

Given their cost and significant impacts on users and society, government appraisal and approval processes involve many parties, including those at the political level, civil servants, external experts, and stakeholder groups. Some roles are more clearly described in the models than others.

The project idea may originate from various sources, including local politicians, stakeholders, and the government. The generation and processing of ideas belong to the informal part of the front-end stage and are, in most cases, not regulated in the model.

Ministries and agencies in respective sectors conduct project appraisal and planning. The same agencies are typically responsible for implementing the projects once approved.

They use tools and guidelines that the model may or may not regulate. Agencies with many projects are typically more experienced and skilled. Quebec stands out by having a single agency responsible for planning and implementing most (but not all) infrastructure projects across sectors.

The QA function follows the ‘three lines of defence’ approach, where quality is first assured by (1) those responsible for the project, then by (2) a separate internal unit, and (3) independent assurance for the largest projects. It is the third line that the models primarily regulate. Independent assurance is organized differently across schemes. Norway, Denmark, and Ireland use external reviewers, while most of the others conduct QA internally within their government administrations. Some use a combination of internal and external reviews. For example, in the United Kingdom, a review panel is selected from the Major Projects Review Group, which consists of experts from both the public and private sectors. In the U.S. Department of Energy, both external and internal independent reviews, as well as independent cost estimates, are required to verify the project baseline and risk profile of its portfolio for large projects.

Scheme management, including QA organization, is typically located at the Treasury or Finance Ministry (Australia, the United Kingdom, Norway, Canada, Quebec, Ireland). Even in the United States, where agencies manage their own schemes, the Office of Management and Budget plays a key role in overseeing budget proposals at the final stage before presenting them to the political level. In two countries, Denmark and the Netherlands, the schemes are managed at the line ministry level. In Australia, the system for the very largest or ‘nationally significant’[1] projects is managed by Infrastructure Australia, the Government’s independent infrastructure advisory organization, while the Assurance Review Unit within the Department of Finance manages the QA for all other projects.

Critical decisions, most notably project approval and funding, are made at the political level. In most countries, this is the Government (Cabinet), while in the smaller (including the Scandinavian) countries, it is elevated to the parliamentary level, probably due to these countries often having minority governments. In the United States, the decisions are jointly made between Congress and the Executive branch. In the Netherlands, the national government proposes a decision together with local governments, but it is the Parliament that has to approve the decision. The Parliament can also decide to amend the proposed decision by the government. Some countries also have advisory panels or commissions that provide independent advice to the government. These commissions normally take a broader, strategic view of investment needs and focus less on individual projects.

In most cases, the project documents are sent to the line ministry, via the Ministry of Finance (or other unit responsible for the model), before independent assurance. The line ministry will normally review the documents before forwarding them. Upon approval, the line ministry typically plays a key role in the follow-up and close-out of the project. However, the models do not always regulate this, focusing mostly on the pre-approval stages.

Figure 15.1 illustrates the organizational chart for a typical project model, which may not be entirely accurate for any specific model.

1 Projects requesting over A$250 million from Federal Government or projects assessed as ‘nationally significant’ by Department of Finance.

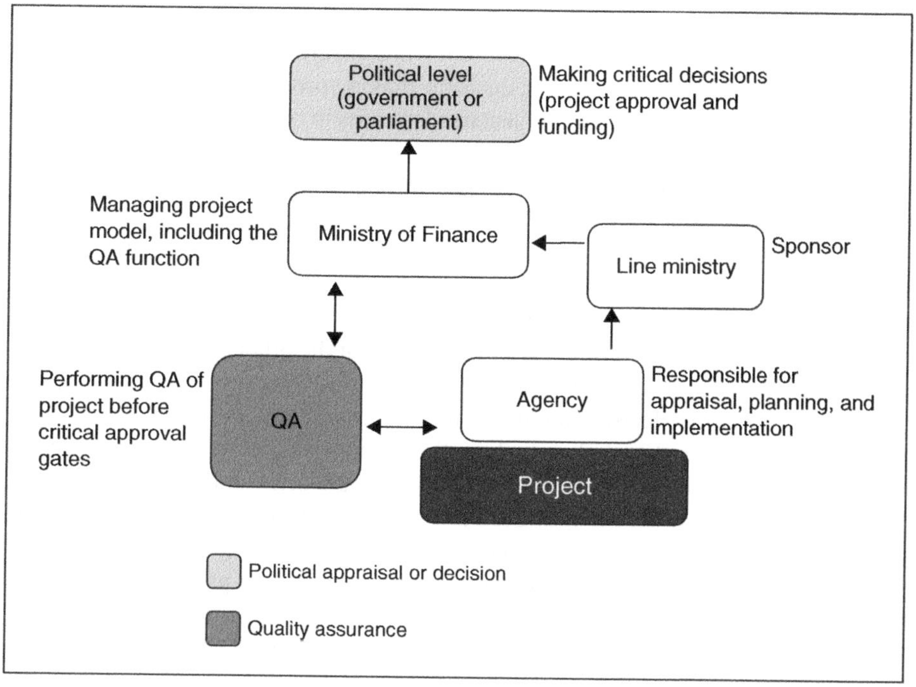

Figure 15.1 Actors and roles involved in a typical project.

15.4.2 Other Parties and Roles

In addition to the parties mentioned earlier, most countries have requirements for involvement and consultation to ensure stakeholder participation. Sweden and the Netherlands have decentralized processes where local and regional authorities explicitly play a role in the model. The process of stakeholder engagement involves layers and complexities that are beyond the scope and mandate of the stage-gate process model, so the focus on stakeholders is secondary in the project models.

Auditors-General, Spending Review commissions, and similar entities may also have a role in scrutinizing investment decisions and controls, typically through one-off reviews after issues have arisen. Such reviews are not part of the schemes but supplementary control functions.

In Norway, a trailing research programme was established to accumulate information about the projects and disseminate good practices. The programme does not have a formal role in the scheme itself but is important in summarizing lessons learned.

15.5 Stage-Gate Processes

A stage-gate process is designed to guide projects from conception to completion through a series of distinct phases, each separated by 'gates'. The models typically require progressively accurate cost estimates to decide whether to move to the next stage and ultimately approve the project's implementation, thereby authorizing project funding. It enhances understanding of the project's purpose and content while ensuring that participants share a common awareness of their position in the decision-making process. A project model may have both mandatory stage-gates that a project needs to pass in order to proceed to the next stage, and stages of a more advisory nature.

The project models presented in this book differ in scope. The number of phases varies from four to six, with stage-gates in between. The extent to which the models also include delivery and operation varies. The emphasis, for most, is on front-end activities, as the content of subsequent stages must be tailored to the specific characteristics of each project. Sweden does not have a project model, but the framework for planning and appraisal of transport projects has much in common with the models of the other countries.

Regarding the number of decision points, Denmark and Norway have the simplest schemes, featuring only two decision points (Denmark has only one for IT projects). These points are limited to the front-end phase, where the main decisions are made, specifically concerning the choice of concept and the project's strategic orientation. The purpose of these decisions is to allow for changes before formal project approval or to terminate the project if necessary. It also varies which body approves the project. In Norway, for example, projects that are part of the scheme must be approved by parliament based on a parliamentary bill, while in Denmark, the largest road, railway, and building projects require a specific law. In other countries, projects are typically approved by the government or by the responsible ministry.

Option appraisal provides a systematic approach to evaluating various choices, ensuring that decisions are evidence- and needs-based. It acknowledges that there are always multiple solutions to a problem. Focusing solely on different versions of the same concept increases the risk of path dependency, which may hinder the identification of efficient solutions and compromise value for money. Significant differences exist regarding the extent to which the models mandate a comprehensive evaluation of options. Most countries require that a reasonable range of conceptually different options is considered in the early stages of project development, while specific requirements to include small or do-minimum options are rare. The Dutch guidelines are interesting because they explicitly state that appraisals should take a broad perspective to avoid tunnel vision. For example, Dutch transport appraisals should not only focus on traffic safety and accessibility but also encompass other dimensions such as sustainability, energy neutrality, housing, landscape, cultural heritage, knowledge of soil and subsoil, and climate adaptation. Sweden offers another intriguing example as it is the country that developed the so-called four-step principle, which requires the analysis of four different types of measures when addressing transport problems: (1) measures that influence transport needs and choices of transport mode, (2) measures that lead to more efficient utilization of the existing road network, (3) minor road improvements, and (4) major investments. While the Swedish project model is sector-specific, the four-step principle aims to ensure that smaller measures lower on the intervention scale, are also examined. However, the Swedish National Audit Office has criticized the transport administration for not fully following the principle, and in particular for not exploring options at step 1.

The extent to which conceptually different alternatives are included in project appraisal, or if the appraisals are effectively centred around a preferred option, is an area that needs to be explored empirically. This illustrates that theory (the models) may differ from practice.

QA can offer an external perspective on projects and help identify and mitigate potential issues early in the project life cycle, reducing the likelihood of defects and errors. All countries utilize elements of QA, but the scope and responsibility for it vary, as discussed earlier. The scope of QA also differs. Norway, Ireland, and Denmark (to some extent) employ external QA, whereas in the other countries, projects are reviewed internally or by other government agencies. The scope of QA also differs. In Norway, the quality assurer must conduct its own independent analyses, while in the other countries, the quality assurer primarily reviews the work that has been completed.

Cost estimation and risk analysis are critical components of project governance regimes, as avoiding cost overruns is crucial to planners and decision-makers. Several countries have developed guidelines for cost estimation within various sectors, but there is no general standard approach to cost estimation. Probability-based estimation is common in the United Kingdom, the United States, Norway, Sweden, and Australia. At the same time, the other countries typically use uplifts to the base cost, which are gradually reduced as the projects mature.

Ex post evaluation is important for promoting learning and accountability, but it is rare in most countries. A standardized governance framework provides an opportunity to institutionalize requirements for evaluation activities from project initiation to delivery and operation. Most of the countries have requirements for monitoring and evaluation to some degree. The assessment of whether a project has delivered its intended objectives is a key part of the schemes in Ireland, Australia, and Canada; however, the actual completion and public availability of these reports are questionable. This lack of transparency undermines public trust and limits opportunities for learning from past projects, potentially reducing the effectiveness of future schemes. Without accessible and thorough reporting, it becomes more difficult to hold stakeholders accountable, draw lessons for policy improvement, and ensure that projects serve the public interest as intended.

Furthermore, actual completion rates (of evaluations) vary, and reports are mostly conducted by the sponsoring agency. Norway has no formal requirements to evaluate projects, but the Concept Research Programme is tasked with evaluating up to four projects per year using a standardized evaluation framework. As such, it provides an external perspective on project success. In Denmark and the Netherlands, projects are required to produce a completion report that details their achievement of goals and realization of benefits. The most comprehensive requirements for evaluating projects can be found in the United Kingdom. Their approval process for projects and programmes requires that all projects and programmes have robust evaluation plans in place. Business cases are unlikely to receive approval without a robust and proportionate evaluation plan that is properly aligned with the delivery context of the specific project or programme. From April 2024, all projects and programmes must log their evaluation plans and outputs in the Evaluation Task Force's Evaluation Registry.

15.6 Evidence of Positive Effects

There is little evidence, in the form of research and evaluations, of the schemes' effectiveness. Some reports from Auditor Generals on projects and practices in selected agencies and sectors exist, but few focus on the impacts of these schemes. There is, however, evidence that gateway review processes have helped ensure projects are evidence- and needs-based and that project planning processes have become more rational and predictable.

Norway stands out with a trailing research programme that follows the scheme and projects. Following the introduction of external assurance of the final budget (QA2), stochastic estimation is used consistently, and cost estimates have become more realistic. The basis for early strategic assessments has also improved, with high-quality problem analyses, option appraisals, and CBAs. Initially, few guidelines and training were available, but this is now in place, and practice is becoming more harmonized. However, there is still a room for improvement in areas such as early cost estimation, treatment of do-minimum options, benefits management, and ex post evaluation.

Some explanations for the positive effects in Norway include the strict management of the scheme (exemptions are not easily obtained) and independent QAs providing ministries and agencies with a motivation to improve their skills and practices. In turn, improved practices in ministries and agencies, supported by full transparency of project documents, may have disciplined politicians, who now have less scope to sidestep the appraisal process and fast-track projects for approval. The trailing research has also helped establish learning loops within government, with research findings used in teaching at universities and in-house training in ministries and agencies.

In the United States, many agencies have adopted stage-gate processes that are used in conjunction with decision support tools, independent internal and external reviews, and a focus on integrated project teams as means to make more informed and risk-based decisions. Earned Value Management Systems (EVMS) are mandatory on projects above certain thresholds (US$50 million), with the systems having to be set up during the front-end planning process to be successful. There is evidence that these efforts have led to more successful results.

Reports from the World Bank show a correlation between project success and the quality of decision documents. In Denmark, improved cost control in road projects has led to a reduction in the mark-up. In the United Kingdom, after a new version of the scheme was introduced in 2011, several evaluations of the workings of the scheme conducted in 2012–2013 concluded that it was appropriate to establish a separate body (Major Projects Authority [MPA],[2] later IPA,[3] and now NISTA[4]) to support more effective management and delivery across government. It was considered particularly important that this body was given a mandate directly by the Prime Minister, whereas the previous version of the scheme was weaker and less formalized.

2 MPA, established in 2011, to replace the OGC.

3 IPA, established 2016 when the MPA was merged with Infrastructure UK.

4 NISTA, established 2025 to combine the functions from the former IPA and the National Infrastructure Commission.

Another indication of the benefits of schemes is their adoption by agencies and projects outside the schemes. In Norway, the Ministry of Finance's project model has been adopted (or variants of it) by some state-owned enterprises and municipalities. In the Netherlands, the MIRT model, managed by the Ministry of Infrastructure and Water Management, has been partly adopted by another ministry. In Denmark, the governance model for transport projects is also voluntarily applied to construction projects.

In Quebec, researchers have demonstrated positive effects on three levels: (1) at the institutional level, the framework increases the legitimacy and credibility of the government; (2) at the organizational level, it strengthens accountability by having a formal organization in charge of project management (SQI) and developing governmental project management capabilities; and (3) at the project level, having to document a project thoroughly helps project teams to be more efficient (Brunet 2021).

15.7 Lingering Issues with Project Governance

It should be said that the intention of all project governance schemes is to improve cost and benefit performance of public investments. However, while there is some evidence of their effectiveness (as discussed in the previous section), there is nonetheless a lack of consistent data across all the case countries to robustly confirm their positive effects. For example, there is still considerable discussion about cost escalation and overruns in several countries, which raises questions about the effectiveness of the project governance schemes. In the United Kingdom, where the Delivery Confidence Assessment ratings for the Government Major Projects Portfolio (GMPP) are presented annually, it has been noted that ratings have not improved in recent years despite assurance reviews. In Denmark, serious problems persist with managing the largest and most complex IT projects. More importantly than cost escalation and time overruns, however, we note that it is the success or failure in delivering benefits that should be taken as a more reliable indicator of the effectiveness of project governance schemes.

At the same time, there are recurring reports in several countries of political interference and early commitments leading to schemes not being fully adhered to. Examples from Australia, Quebec, Sweden, and the United Kingdom demonstrate that politically initiated projects can be treated as exceptions to the governance process (that is, the business case is used as a checklist for preparing a funding request rather than as an instrument to support rational decision-making). There are also numerous reported instances of notable gaps in the completion of ex post evaluations, which prevent learning from previous projects. Organization and management can also be challenging. In some countries, delegating responsibilities from the line ministry to an agency and, ultimately, the project manager remains unclear (even though responsibilities may be clearly defined on paper). A lack of follow-up and evaluation during implementation and after delivery is often seen as a weak link.

Lack of transparency is a related issue. In Australia, there is a broad consensus on the value of transparency in general; however, governments rarely release the advice they use in making project decisions. Post-project reviews by Auditor-Generals or other independent parties are openly available but are conducted too late to correct poorly informed decisions. By contrast, in countries like Norway and Denmark, where project appraisals, review reports, and ex post evaluations are openly available, this will contribute to accountability on the part of civil servants, quality assurers, and decision-makers. However, it should be noted that releasing reviews and evaluations may be a double-edged sword, as public release of data can lead to excessive editing of the content and the use of risk-averse language. This would be a legitimate argument for not disclosing the review reports to the public in all cases.

Organizational capability gaps pose problems, and capability development programmes (within areas relevant to project management and governance) are not a key element of most schemes; this segment is left to agencies and to universities. For governments to implement capability development initiatives, there needs to be a baseline of capability in the organization. In project-based settings, the capability is difficult to retain as individuals leave for other projects and have no interest in remaining with the government organization and help deliver the upcoming stream of investment projects. As a result, the implementation of capability development programmes has been rather patchy across the different countries. Some good examples include the Major Projects Leadership Academy (MPLA) in the United Kingdom, aimed at civil servants within the project delivery profession in the public sector, the Project Leadership Institute run by the U.S. Department of Energy to prepare project leaders for complex government energy projects, and the Australian Major Project Leadership Academy for Australian public sector project leaders. Similarly, Sweden's transport authorities have launched a programme to improve their cost estimation capacity. Ireland's relatively recent project governance scheme notes the need to strengthen competence among those involved in evaluating, planning, and implementing public sector projects. On balance however, it should be acknowledged that there is not much uptake of capability development programmes across the countries studied. As a result, there is an opportunity to develop training and mentorship programmes within governments for project governance capability. While research-intensive academic education providers seem uniquely positioned to provide such capability development programmes, this does not preclude organizations themselves from developing less formalized in-house programmes based on peer learning and knowledge transfer.

15.8 Schemes May be Contested

There is dissatisfaction with the schemes in several countries. The most serious concern is that the models are rigid and inflexible, prolonging planning times (and possibly increasing costs) for large public projects, without necessarily improving their outcomes. Similarly, in a political climate that encourages accelerated decision-making and streamlines the process, there is often criticism that such governance schemes can stifle many innovative and transformative ideas due to a bureaucratization of the decision-making process and emphasis of risk-averse behaviour. Another criticism of the schemes is that since many projects have been politically endorsed and de facto selected in advance, the appraisal and planning process becomes an exercise in validation and a mere rubber-stamp process. Even without the political dimension, it may seem obvious that a certain project is needed, making investigating alternatives seem worthless. In some countries, it has been argued that there should be wider possibilities to skip certain steps when time is critical.

Critics of the models have also argued that they should be better adjusted to the mode selected. For example, projects with collaborative delivery modes should not have to go through the same restrictive procedures. Similarly, it has been argued that schemes are not adapted to IT projects using an agile approach. In some countries, IT projects follow a different model, but it is still normally some stage-gate model or hybrid model.

Another concern relates to sustainability considerations, particularly climate policy, which have not yet been given a serious position in most models.

15.9 Summary of Experiences

- The quality of decision documents has improved.
- Processes have become more streamlined.
- Cost control has improved in some countries, but challenges remain, and early estimates are still biased.
- The schemes may have a disciplining effect on political decisions, but the tension between politics and rational planning and decision-making is a perennial issue.
- Whether the frameworks work according to their intention may depend on transparency, skills development, and whether the scheme is mandatory.
- Over-bureaucratization is a risk – schemes are seen as ‘red tape’.
- The overarching process is a socio-technical system, and hence very complex, with several potential modes of failure.

In summary, while governance schemes have contributed to improved documentation quality, more streamlined processes, and, in some cases, better cost control, they are not without significant challenges. The tension between political priorities and rational planning remains unresolved, and the risk of over-bureaucratization threatens to undermine the very efficiency these models aim to promote. For these frameworks to truly support better decision-making, they must evolve to accommodate different project types, delivery modes, and emerging priorities such as sustainability and agility. A more flexible, context-sensitive approach – one that balances structure with adaptability – may be key to unlocking their full potential and ensuring they remain relevant in an increasingly complex and fast-paced planning environment.

Reference

Brunet, M. (2021). Making sense of a governance framework for megaprojects: the challenge of finding equilibrium. *International Journal of Project Management* 39 (4): 406–416.

16

Conclusions

Vedran Zerjav, Morten Welde and Gro Holst Volden

Concept Research Programme, Norwegian University of Science and Technology (NTNU), Trondheim, Norway

This concluding chapter synthesizes insights from the preceding chapters. It reflects on the role of structured decision-making models, particularly stage-gate processes, in enhancing the legitimacy, accountability, and efficiency of public investments. Drawing on theoretical perspectives and country-level experiences, this chapter explores how governance frameworks can reduce uncertainty, delay premature commitments, and foster transparency and trust. It also highlights the importance of learning, adaptation, and long-term outcomes, offering guidance for policymakers seeking to strengthen public sector project governance in an increasingly complex and dynamic environment.

16.1 What Do We Learn and Where Do We Go from Here?

According to Brunet and Aubry (2016) and Brunet (2021), the main roles of a project governance framework are to increase government legitimacy, accountability, and efficiency (see Figure 2.1). Arguably, the best way to support and bring coherence and legitimacy to the decision-making process is to formulate it as a structured and collective process with provisions for including multiple voices and points of assessment. Using a governance system and a project model brings clarity and procedural fairness to the fuzzy decision-making process, which is otherwise often unreliable. The main position in this book is, therefore, that we see a project governance framework as an artefact of robust decision-making on public and fiscal spending. Country cases confirm that, for the most part, a stage-gate project governance model can provide a structure and create conditions for transparency to lead to more legitimate decisions and implementation pathways. As such, they provide a solid basis for collective decision-making on public spending in mature representative democracies. However, for a governance system to be effective, the models that underpin it also need to be transparent. Most (if not all) of the country examples and settings provide relatively straightforward models, which can be understood by most non-experts/generalist audiences.

The main question we seek to answer in this part of the book is to what extent the use of a project governance framework supports the legitimacy, accountability, and efficiency

Governance Frameworks for Major Public Projects: International Practices and Experiences, First Edition. Edited by Vedran Zerjav, Morten Welde and Gro Holst Volden.

of government decisions (compared with the situation where decisions are implemented without such models)?

In seeking to uncover the diverse practices and experiences, we selected a sample of mature 'Western' democracies as country case settings. Having briefly described and compared the frameworks in those settings, we arrive at the fundamental question for governments and policy makers: What is the value of having a normative stage-gate process model in relation to decision-making robustness?

The extent to which a project governance framework is useful in its application is the access and availability of the data to be used in the model and for ex post evaluation. Moreover, a project model should be based on implementing a specific theory of change that defines the purpose and logic of the intervention. While large-scale project investments are often discussed in terms of their positive externalities, neither positive nor negative externalities nor secondary effects should be relevant in the governance model. By contrast, the governance model should ensure that the project stays true to its mission and theory of change, leading to its designed outcomes (rather than serendipitous or secondary effects).

Emphasis of the state project models is on the front end of projects (before projects get sanctioned) because it is where decisions with the most impact are taken. It is also the early phases of projects (upstream of policy formulation and definition) where the governance frameworks and systems are mainly implemented. This confirms the well-established position that decisions made in the early front end of project implementation (or the upstream of policy definition) count most in shaping project outcomes. However, we extend this notion with a simple pragmatic insight that while the front end governance is essential, it cannot turn around a fundametally misguided project or eliminate the need for adaptation in downstream project implementation. Decision-makers should, therefore, ask both whether there has been enough focus on the front end and whether there has perhaps been too much focus on the front end.

Alongside the more established discussions on the front end, there is much less emphasis on the back end, the phase when the project is handed over and needs to start delivering value to its stakeholders, users, and other beneficiaries (often the general public). For example, Canada uses the reference period of 6–18 months for ex post evaluations, and there can be a discussion as to how long should the reference period be to capture all the relevant benefits and outcomes. In most countries included in this overview, outcomes are not consistently captured, making it difficult to ascertain whether projects that used the governance system fulfilled the long-term goals and benefits stipulated in their strategic or political brief. Similarly, given the limited experience with using the models, it should not be deduced that a normative stage-gate process model automatically increases the legitimacy, efficiency, and accountability of the government decisions being made. It is important to know whether the costs of designing, operating, and running such a governance system justify its outcomes.

Favourable outcomes in governance include decisions that would lead to better projects in terms of tactical delivery performance (lower cost overruns and delays) and strategic mid-term outcomes (committing to the right projects). To be legitimate, the decision-making process should also have procedures that make it robust and capture the mid- and long-term benefits and co-benefits. Based on this, and our review of the country examples, we suggest that a project governance system is effective if it

(1) reduces uncertainty, (2) delays commitment to a point of lower uncertainty, (3) challenges assumptions on causality and theory of change, (4) provides an artefact for building transparency and trust, and (5) links the decision-making process to long-term project outcomes. We next discuss these goals of an effective project governance framework with reference to the three governance theoretical angles introduced in Chapter 2.

First, a project governance framework must reduce uncertainty. Reducing uncertainty lowers transaction costs, which are often amplified when information is incomplete or dispersed. Lower uncertainty also suggests that decision-makers operate on a shared understanding of the project environment and the expectations of the project, reinforcing trust among stakeholders.

Second, a project governance system should delay major commitments until uncertainty has been sufficiently reduced. Commitments to large-scale investments are typically irreversible and carry significant lock-in and sunk cost risks. A staged approach to decisions and approvals, supported by evidence thresholds, allows decision-makers to build their confidence levels in the quality of the project, thus reducing the lock-in and sunk cost risks of a potentially premature commitment.

Third, a project governance system must challenge assumptions about causality and the project's theory of change. Projects can fail due to a vaguely defined value proposition and untested initial assumptions about the problem situation they are addressing. By requiring explicit articulation of benefit hypotheses (i.e. as a theory of change) and utilizing mechanisms for periodic validation, project governance should ensure that decisions are grounded in evidence rather than biased information.

Fourth, a project governance system is an artefact that focuses attention, catalyses interactions, and acts as a source of transparency and trust. The stage-gate model, however simplified, makes the decision-making process visible and understandable to stakeholders. As an artefact, the stage-gate model enhances procedural fairness and reduces perception of arbitrariness in decisions being made. This visibility not only lowers the costs of monitoring but also signals a commitment to shared purpose, reinforcing the legitimacy of both the process and its outcomes.

Finally, a project governance model must link the decision-making process to long-term project outcomes and maintain continuity from front-end early decisions to the realization of the project's long-term strategic objectives. This continuity is crucial for ensuring that projects remain relevant and deliver value to stakeholders and society as a whole. The long-term focus on project governance ensures that decisions made today remain defensible in the future.

In this way, an effective project governance model, through its transparency and clarity, provides a foundation for greater accountability, efficiency, and legitimacy of government decisions in complex project contexts.

We further discuss two additional aspects of the application for the models: (1) decision-making and optimization level and (2) timing and scaling.

The first consideration is at what level we should seek the impacts of the governance model. While the decisions are being made about the project, the project's outcomes are only relevant in relation to the policies being implemented. Those policies are often at the sectoral level (transport, information and communications technology [ICT], health, defence, etc.) and are translated into projects. This relates to one of the most long-standing discussions in the project management research community: whether to prioritize process efficiency (project management success) over the project's success, seen as an outcome of this process (Bosch-Rekveldt et al. 2023), or vice versa. Similarly, in policy studies, there has been a discussion on the success of policy through its multiple streams – the problem, solution, politics, process, and programme (Howlett 2019). However, since post hoc evaluation data is either incomplete or missing, it is challenging to directly relate the project's effectiveness to the policies being implemented (Samset and Volden 2022). Even when such evaluations are undertaken, there is seldom a political incentive to publish them. Moreover, there is very limited evidence of the use of portfolio-level governance (how individual projects create a balance concerning the entire set of activities in the sector or even across sectors). The closest example is perhaps the Dutch Regional Development Agenda, where geographic regions are viewed as a portfolio of projects for planning purposes. While it is to be expected that other countries run a similar model at the level of planning, this is still relatively far from the conventional notions of portfolio management.

This brings us to ideas of the schemes' timing and scaling. The desired impact is about the extent to which a governance model guards against lock-in and escalation of commitment before sufficient details about the project and its outcomes are known. Timing is a key aspect, so it is, for example, important that the cost–benefit analysis (CBA) and quality assurance (QA) take place before the escalation of commitment and feed into the political decision-making, as opposed to the QA process being a box-ticking exercise to legitimize pre-existing political decisions. This is related to the political resistance that is sometimes observed in relation to the use of transparent governance frameworks and the institutions put in place to manage this resistance. For example, in Norway, the key mechanisms were the strong role of the Ministry of Finance (MoF), the strategy to start small (QA2, which was less controversial) and then expand the scheme.

Ultimately, the efficiency of a governance system is determined by its ability to evolve, adapt, and remain relevant over time. Governments should, therefore, invest in building organizational capabilities to embed dynamic adaptation and learning into organizations that run the project governance scheme, ensuring that project models are resilient and future-oriented.

It is the key advantage and limitation of project governance stage-gate models that they usually treat project phases as linear and discrete. It is an advantage because it helps generate coherent and easy-to-understand artefacts to centre communication and decision-making around. It is also a limitation because it can constrain the flexibility of the model to absorb developments and changes and recalibrate project goals and key stages, as the need arises. This requires embedding feedback loops not only at the back-end (ex post evaluation) but also throughout the project life cycle, including mid-term reviews and real-time performance monitoring.

Learning mechanisms should be institutionalized within the governance framework to ensure that insights from past projects inform future decision-making. This includes:

- **Systematic capture of lessons learned**, both from successful and failed projects.
- **Cross-sectoral knowledge sharing, enabling transfer of practices across domains (e.g. health to transport to ICT)**.

- **Stakeholder engagement, where users and beneficiaries are brought together and can provide meaningful input on project relevance and outcomes.**
- **Dynamic updating of programme theories of change, reflecting evolving societal needs and technological possibilities.**

This requires data infrastructure that enables evidence-driven decision-making and integration of project data repositories to enhance transparency and responsiveness to changes and enhance nuance in the data. While this is a major opportunity for improving project governance schemes, it also raises questions about data governance, privacy, and the capacity of institutions to interpret and act on complex datasets.

16.2 Project Governance Models and Governance Theory

We finally discuss the project governance frameworks – particularly the structured, stage-gate models – in relation to the theoretical lenses introduced in Chapter 2 of the book that introduced Transaction Cost Economics (TCE), Agency Theory (AT), and Stewardship Theory (ST) as main theoretical angles to interpret project-level governance. Each theory brings into focus a different facet of how decisions are made, justified, and sustained over time.

From one angle, the models reflect a concern with efficiency and risk mitigation, echoing the logic of TCE. The emphasis on delaying commitment until uncertainty is reduced, and on procedural clarity to avoid costly renegotiations, speaks directly to the need to manage complexity in public investment decisions. These models are not just about reducing costs – they are about creating institutional scaffolding that makes decision-making more predictable and less vulnerable to opportunism.

At the same time, the governance schemes focus on both accountability and control, themes central to AT. The structured checkpoints around stage-gates, QA mechanisms, and ex post evaluations are all designed to ensure that projects deliver value to their stakeholders, and the broader public and that they remain aligned with policy objectives. Yet, the experience with models also reveals the limits of control – especially when political incentives override technical assessments and analysis, which calls for a more nuanced understanding of how and when accountability mechanisms should be applied.

But most compelling is the way these models open space for a stewardship perspective. By anchoring projects in a theory of change suggesting how public value will be created and by emphasizing transparency and collective rationality, the project governance models invite actors to act not just as agents under scrutiny but also as stewards of public purpose. Bringing learning, adaptation, and long-term outcomes into focus, the country cases suggest that governance is not merely a system of checks and balances – it is a platform for trust, collaboration, and shared responsibility. While there are elements of the stewardship approach in all of the country case examples, they are, perhaps more prominent in the Scandinavian countries, arguably due to higher levels of trust in public institutions and systems and flatter organizational structures, which allow for less preoccupation with individual accountability for successes and failures. Nonetheless, when project governance models are viewed as artefacts for transparent and structured collective decision-making in the public sector context, it is clear that the foundation of ST underpins the logic of all of them.

No single theoretical construct can fully capture the complexity of governance and its decision-making, but together, they offer a more complete picture of what our analysis of the project governance models suggests.

16.3 Advice for Policymakers

Project governance models may play a crucial role in promoting project success and thus fiscal efficiency in a world largely characterized by historically large and increasing public debt levels. Gateway review processes are useful for ensuring decision-makers receive high-quality analyses at appropriate stages, thereby enhancing projects' efficiency and effectiveness. More generally, they can also strengthen the accountability and legitimacy of public projects.

Consistent with textbooks on project management and governance, early strategic appraisals should be based on problem analysis, broad option assessments (such as the 'four-step principle'), and CBAs. Goals for users and society should be rooted in a sensible theory of change. As planning becomes more detailed, estimates of costs and benefits must rely on sound methods (such as stochastic analysis) and empirical data and experiences. Additionally, active management of costs and benefits should be maintained throughout the project life cycle and after completion. A good balance between robust planning and flexibility to accommodate changes must be achieved. However, while decisions made in the front end are important, no amount of planning can turn a wrong project into the right one.

The schemes presented in this book all provide a structured approach to project appraisal and development. It is beyond the scope of the book to evaluate the success of the models in their respective countries. However, the models have some common features that suggest that successful project governance schemes should include:

1) **Strict management of the scheme:** Although there should be some flexibility, as projects differ in their need for governance, the scope to sidestep the process should be limited. A lack of consistency in following the process from project to project may induce a 'uniqueness bias', whereby project promoters may undervalue the potential for learning from past experiences and similar projects. There must be an authority with clear responsibility for managing the model, and only this authority should be allowed to grant exemptions.
2) **Independent QA:** The use of (internal and/or external) QA for decision documents is essential to prevent the optimism bias commonly seen in public projects.
3) **Strong involvement by the Ministry of Finance or Treasury:** This can inject fiscal realism at the early stages of planning and ensure consistency in QA across different sectors.
4) **Clear decision hierarchy:** A decision hierarchy with clear roles and responsibilities for the project is essential. Senior Responsible Owner (SRO), or a similar position in the line ministry, should play a key role in the governance system.
5) **Gradual implementation:** Governance models may be unpopular with decision-makers because they can hinder project initiation. Many countries with comprehensive

schemes today initially started with a simple model, which was later expanded, such as by increasing decision points and coverage of sectors. This approach could be a practical option for countries lacking governance models.

6) **Transparency:** Making project appraisals and evaluations public can foster improved practices and strengthen the democratic debate on projects. However, an over-emphasis on transparency might also be detrimental and result in excessively risk-averse project participants. A proper balance between these factors should be maintained.
7) **Capability and competence development:** Competence is essential for all participants in projects, including SROs. Although a QA system with external experts is beneficial, it cannot replace the expertise of skilled public planners. Different stages of a project may demand varying types of competence and tools. Training can be provided internally or in partnership with academia.

There are also several pitfalls to steer clear of:

1) **Inconsistent project adherence:** A lack of consistency in applying the project development process across different initiatives can undermine the credibility and effectiveness of the overall model. When the process appears to be followed selectively, it risks being perceived as a theoretical framework that looks good on paper but lacks practical enforcement or value.
2) **Under-resourced front-end phase:** The front-end planning stage is often inadequately supported with funding and personnel. This lack of investment can compromise the quality of early decision-making. Given that front-end loading is critical to project success, it is essential to allocate sufficient resources to this phase.
3) **Delayed strategic appraisal:** Strategic direction should be established early to ensure alignment between projects and broader societal goals. Conducting strategic appraisals after implicit commitments have been made reduces their effectiveness and limits the ability to influence outcomes meaningfully.
4) **Insufficient option appraisal:** Projects often focus on a single preferred solution without thoroughly exploring alternative approaches. This narrow scope can lead to missed opportunities, especially when infrastructure investments are pursued without confirming they are the most suitable response to the problem.
5) **Lack of project maturity:** Advancing through decision gates before a project is ready – under the assumption that issues can be resolved later – can result in significant downstream problems. While early-stage uncertainty is natural, the goal of early appraisal is to identify and manage risks proactively.
6) **Overcomplication:** Effective project models should be high-level and focused on the key issues necessary for political and strategic decision-making. When models become overly detailed and time-consuming, they risk being perceived as bureaucratic 'red tape' and may lose their relevance and influence.
7) **Insufficient follow-up and learning:** Project appraisals and decisions are typically based on forecasts and assumptions about future outcomes. However, there is often limited knowledge about how projects actually perform relative to their initial goals. Without systematic ex post evaluations of completed projects, valuable learning opportunities are lost, and the potential to refine and improve project models over time is significantly diminished.

8) **Stakeholder involvement imbalance:** Effective stakeholder engagement is essential for legitimacy and informed decision-making. However, both over-involvement and under-involvement can be problematic. Excessive engagement may lead to inflated expectations and politicization of the process, while insufficient involvement can result in a lack of buy-in and missed insights from key actors.

Tension between political considerations and rational planning is an inherent feature of large, government-funded projects. These projects are deeply political by nature, which can lead to narrow perspectives, prolonged decision-making processes, and outcomes that may appear unpredictable from a purely analytical standpoint. However, as Samset and Volden (2016) point out, the tension between political and technical processes is not a flaw but a fundamental aspect of democratic governance. Therefore project governance models should not be seen as technocratic systems, which offer efficiency but can lack legitimacy. They should be seen as artefacts of democratic processes that ensure broader accountability and public trust. In this way, the ultimate goal of project governance models is to strengthen the analytical foundation of decision-making and strive to ensure that political leaders are well-informed and guided by robust evidence.

References

Bosch-Rekveldt, M., Bourne, M., Forster, R. et al. (2023). Performance measurement in project management. *International Journal of Project Management* 41 (7): 102518.

Brunet, M. (2021). Making sense of a governance framework for megaprojects: the challenge of finding equilibrium. *International Journal of Project Management* 39 (4): 406–416.

Brunet, M. and Aubry, M. (2016). The three dimensions of a governance framework for major public projects. *International Journal of Project Management* 34: 1596–1607.

Howlett, M. (2019). Moving policy implementation theory forward: a multiple streams/critical juncture approach. *Public Policy and Administration* 34 (4): 405–430.

Samset, K. and Volden, G.H. (2016). Front-end definition of projects: ten paradoxes and some reflections regarding project management and project governance. *International Journal of Project Management* 34 (2): 297–313.

Samset, K. and Volden, G.H. (2022). Closing the loop: ex ante and ex post evaluation in order to learn from mistakes and successes. In: *The Front-End of Large Public Projects: Paradoxes and Ways Ahead* (ed. T.M. Williams, K. Samset, and G.H. Volden), 158–190. Abingdon: Routledge.

Index

Governance Frameworks for Major Public Projects: International Practices and Experiences, First Edition.
Edited by Vedran Zerjav, Morten Welde and Gro Holst Volden.

d

t

Printed and bound by CPI Group (UK) Ltd, Croydon, CR0 4YY
17/09/2026
14964270-0001